Contributions of the Committee on Desert and Arid Zones
Research
of the Southwestern and Rocky Mountain Division of the
American Association for the Advancement of Science

Previous Symposia of the Series

1. *Climate and Man in the Southwest*. University of Arizona, Tucson, Arizona. Terah L. Smiley, editor. 1957
2. *Bioecology of the Arid and Semiarid Lands of the Southwest*. New Mexico Highlands University, Las Vegas, New Mexico. Lora M. Shields and J. Linton Gardner, editors. 1958
3. *Agricultural Problems in Arid and Semiarid Environments*. University of Wyoming, Laramie, Wyoming. Alan A. Bettle, editor. 1959
4. *Water Yield in Relation to Environment in the Southwestern United States*. Sul Ross College, Alpine, Texas. Barton H. Warnock and J. Linton Gardner, editors. 1960
5. *Ecology of Groundwater in the Southwestern United States*. Arizona State University, Tempe, Arizona. Joel E. Fletcher, editors. 1961
6. *Water Improvement*. American Association for the Advancement of Science, Denver, Colorado. J. A. Schufle and Joel E. Fletcher, editors. 1961
7. *Indian and Spanish-Ameican Adjustments to Arid and Semiarid Environments*. Texas Technological College, Lubbock, Texas. Clark S. Knowlton, editor. 1964
8. *Native Plants and Animals as Resources in Arid Land of the Southwestern United States*. Arizona State College, Flagstaff, Arizona. Gordon L. Bender, editor. 1965
9. *Social Research in North American Moisture-Deficient Regions*. New Mexico State University, Las Cruces, New Mexico. John W. Bennett, editor. 1966
10. *Water Supplies for Arid Regions*. University of Arizona, Tucson, Arizona. J. Linton Gardner and Lloyd E. Myers, editors. 1967
11. *International Water Law Along the Mexican-American Border*. University of Texas at El Paso, El Paso, Texas. Clark S. Knowlton, editor. 1968
12. *Future Environments of Arid Regions of the Southwest*. Colorado College, Colorado Springs, Colorado. Gordon L. Bender, editor. 1969
13. *Saline Water*. American Association for the Advancement of Science, New Mexico Highlands University, Las Vegas, New Mexico, Richard B. Mattox, editor. 1970
14. *Health Related Problems in Arid lands*. Arizona State University, Tempe, Arizona, M. L. Riedesel, editor. 1971
15. *The High Plains: Problems in a Semiarid Environment*. Colorado State University, Fort Collins, Colorado. Donald D. MacPhail, editor. 1972
16. *Responses to the Dilemma: Environmental Quality vs. Economic Development*. Texas Tech University, Lubbock, Texas, William A. Dick-Peddie, editor. 1973
17. *The Reclamation of Disturbed Arid Lands*. American Association for the Advancement of Science, Denver, Colorado, Robert A. Wright, editor. 1978.

Energy Resource Recovery in Arid Lands

**The Committee on Desert and Arid
Zones Research
of the Southwestern and Rocky Mountain Division
of the
American Association for the Advancement
of Science
Statement of Purpose**

The objective of the Committee on Desert and Arid Zones Research is to encourage the study of phenomena relating to and affected by human occupation of arid and semiarid regions, primarily within the areas represented by the Southwestern and Rocky Mountain Division of the A.A.A.S. This goal involves educational and research activities, both fundamental and applied, that may further understanding and efficient use of our arid lands.

COMMITTEE

Chairman:
 Robert A. Wright, West Texas State University, Canyon
Secretary:
 John A. Ludwig, New Mexico State University, Las Cruces
Members:
 Joseph R. Goodin, Texas Tech University, Lubbock
 David K. Northington, Texas Tech University, Lubbock
 James W. O'Leary, University of Arizona, Tucson
 Marvin L. Riedesel, University of New Mexico, Albuquerque
 Klaus D. Timmerhaus, University of Colorado, Boulder

Mailing Address
Joseph A. Schufle, Executive Secretary
Southwestern and Rocky Mountain Division of A.A.A.S.
New Mexico Highlands University
Las Vegas, New Mexico

Energy Resource Recovery in Arid Lands

Edited by

Klaus D. Timmerhaus

University of New Mexico Press

Albuquerque

Library of Congress Cataloging in Publication Data

Main entry under title:

Energy resource recovery in arid lands.

(Contribution of the Committee on Desert and Arid Zones Research of the Southwestern and Rocky Mountain Division of the American Association for the Advancement of Science)

Includes bibliographic references
Contents: Preface / Klaus D. Timmerhaus—Energy resources located in arid lands / William H. Dresher—Social, economic, and cultural impacts of energy resource development on Indian lands / Thomas Leubben—[epic]

1. Power resources—West (U.S.)—Addresses, essays, lectures.   2. Energy development—West (U.S.)—Addresses, essays, lectures.   3. Arid lands—West (U.S.)—Addresses, essays, lectures.   I. Timmerhaus, Klaus D.   II. Series

T.J.163.25.U.6E54   333.79′0978   80-54573
ISBN 0-8263-0564-4

Contents

Preface

Klaus D. Timmerhaus
University of Colorado
Boulder, Colorado

In recent years it has become apparent that a large share of this nation's future energy must come from the Western States and particularly from arid and semiarid regions of these states. These arid lands offer a variety of energy resource recovery opportunities; however, the recovery of these resources promises to require considerable planning and investment if this development is to be performed in an ecologically and sociologically acceptable manner. To more fully examine these concerns, the Committee on Desert and Arid Zones Research of the Southwestern and Rocky Mountain Division of the American Association for the Advancement of Science sponsored a symposium at Fort Lewis College in Durango, Colorado on April 26-27, 1979.

A large fraction of the energy resources located in arid regions of the United States is controlled by the federal government or is within the boundaries of Indian reservations. Federal control of the lands makes energy resource development considerably more difficult than under private control. Energy resource development on reservation lands adds to these complexities because it requires the potential resource developer to understand the dilemma which Indian tribes face as they attempt to strike a balance between the development of these resources for their own benefit and the preservation of their ancient traditions. Too often, decision making necessitated by resource development has led to unreconcilable internal tribal conflict.

Recovery of energy resources particularly in arid and semiarid regions often also involves significant environmental disturbance. Even though the American public is concerned about their energy supply, they are presently still more concerned about their environment. Thus, there have been and will continue to be numerous philosophical conflicts between energy development and environmental protection. It is, therefore, not surprising that all phases of energy development are rather extensively controlled by federal, state or local statutory regulations. Unfortunately, some of these regulations and requirements have, at times, essentially blocked all efforts to economically recover the needed energy resources. Additionally, they have done little to foster research and development of new energy alternatives.

When energy resource recovery in arid lands does become a reality it will be a major undertaking involving complex processing that requires minimum amounts of water. Since these energy recovery developments will be considerably removed from major population centers, an elaborate infrastructure will be required with relatively large financial commitments. However, it appears, because of the various risks involved, that incentives may have to be developed if such financial support is to be generated. Such incentives could include relaxation or clearance of environmental regulations, streamlining of the complex permit system, expeditious leasing of federal mineral lands, improved depreciation rates, and accelerated tax write-offs.

The eight papers in this monograph briefly address various aspects of these critical problems associated with energy development in arid and semiarid lands. It is hoped that these discussions will provide a clearer perspective of the complexity of these problems and will lead to further work directed toward their solution.

1

ENERGY RESOURCES LOCATED IN ARID LANDS

William H. Dresher

College of Mines
University of Arizona
Tucson, Arizona

SUMMARY

For a number of years a major fraction of the world's energy has been derived from arid lands. So long as those arid lands were in a foreign country and the energy derived was in the form of petroleum, there was little concern for the potential impact of energy resource development in arid lands. Today, however, when it is becoming more and more apparent that a large share of the energy requirements of the United States must be derived from the Western States and the arid and semiarid regions of these states in particular, concern is growing. While the American public is concerned about their energy supply, the most recent poll conducted by Resources for the Future showed that they are even more concerned about their environment. The arid lands of the United States offer a variety of energy resource opportunities; however, the development of these resources promises to be difficult and expensive if this development is to be performed in an ecologically and sociologically sound manner.

INTRODUCTION

Before discussing the estimates of the energy resources of the arid and semiarid lands of the western part of the United States, it may be helpful to review the present energy mix on which the United States is dependent. For convenience in moving from one energy resource to another and, inasmuch as our major consumption is petroleum, I shall express all energy units in terms of millions of barrels of oil equivalent. In energy terms, one million barrels of oil are equivalent to 5.5 trillion Btu, 1.7 billion KWh of electricity, 5.7 billion cubic feet of natural gas, or 0.23 million short tons of coal.

In 1978 the United States consumed 39 million barrels of oil equivalent a day in total energy. Of this, 48 percent was in the form of petroleum, 27 percent as natural gas, 18 percent as coal, 3 percent as nuclear, and 4 percent as hydro, geothermal, and solar. This is equivalent to a total consumption for the year of 14.2 *billion* barrels of oil equivalent or 77 quads (1 quad = 1 billion Btu of energy). Of this amount 9 million barrels of oil per day, or a total of 3.3 billion barrels of oil were imported. This was 22 percent of our total energy supply.

ENERGY RESOURCES OF THE ARID LANDS

While there is no way of determining how much energy is contained in the fossil fuel and natural energy resources of the U.S. arid lands, estimates are that approximately 47 percent of the nation's entire energy resource lies in the western part of the conterminous United States (that is, all western states excluding Alaska)[1] and that this resource could be as much as 3 to 6 times as much energy as that contained in the Middle Eastern countries' petroleum resource.[2] These resources lie in lands which are predominantly arid or semiarid and a large portion of them lie in lands controlled by the federal government. These are the so-called public lands or federal lands of the United States in which the lands are either totally controlled by the federal government or just the mineral resources beneath the lands are controlled by the federal government. As I will mention later, federal control of the lands makes energy resource development appreciably more difficult than if the lands were under private or state control. These energy resources range from the conventional fuels of petroleum, natural gas or coal to the less conventional fuels of oil shale, tar sands, and nonfuel energy resources of geothermal and hydro.

Coal. By far the most predominant energy resource in the West is coal. Nearly half of the U.S. coal identified resource by weight, or over 1 trillion tons, are located in the West and these resources include nearly all of the nation's low (less than 1%) sulfur coal.[3] This coal resource represents over 3 trillion barrels of oil equivalent or approximately 400 years supply of oil-equivalent energy at 1978 levels of petroleum usage! Because the deep beds of western coal are located near the surface of the ground, it is cheaper to produce than is eastern coal. However, it should be noted that western coal contains only about 75 percent of the energy value per ton of eastern coal and therefore there is some offset of this low mining cost. The chief attribute of western coal is its low sulfur content which reduces the cost of air pollution abatement at the power plant where it is burned. Approximately 40 percent of the total U.S. coal resource is under federal lands; 70 to 80 percent of this federal coal is in the western states.[4]

Oil Shale. Oil shale is the next largest energy resource of the western arid and semiarid lands with an estimated 0.4-1.3 trillion barrels of oil contained. These resources occur in the states of Colorado, Utah and Wyoming and could represent as much as 3 to 4 times the proven reserves of the Middle East.[5] Between 70 and 80 percent of all U.S. oil shale deposits are on federally controlled land.

Uranium. Over 90 percent of the U.S. total uranium reserve is located in the western states. Estimates of reserves run between 690 to 920 thousand tons of U_3O_8 (1979 estimate at $30 and $50 per pound).[6] If fissioned in a light water reactor, this amount would be equivalent to between 60 and 90 billion barrels of oil in energy value. The total estimated uranium potential resource of the western states is 4.1 million tons of U_3O_8,[7] an equivalent of over 350 billion barrels of oil. Much of this resource is located in the arid and semiarid regions of New Mexico (52%) and Wyoming (31%) with the remainder in Texas (5%) and Utah, Colorado, and Arizona (9%). An estimated 54 percent of all uranium reserves are located on federal lands in these states.[8]

Petroleum and natural gas. Oil and gas reserve estimates are an issue of great controversy inasmuch as these minerals are much more difficult and costly to assess than coal, oil shale, or uranium. Recent discoveries in the Overthrust Belt region of Wyoming, Idaho, Utah, and Colorado as a result of exploration based on recent interpretation of the modern geologic theory of plate tectonics have greatly increased the enthusiasm of exploration companies. Some fourteen new oil and gas fields have been located. Proven reserves of natural gas at Anschutz Ranch currently totals in the 150 to 200 billion cubic foot range (Keith 1979). The discoveries in this area alone constitute between 8 and 20 percent of the entire undiscovered potential natural gas reserve of the conterminous U.S.! In Arizona, which has not traditionally been an oil and gas producing state, over 4.3 million acres are currently under lease for oil and gas exploration in a large scale test of the extension of the Overthrust Belt into Arizona and southward into Mexico (Keith 1979). Figure 1.1 shows the present interpretation of the Overthrust Belt. Estimates made in 1973-1974, at the time of the Arab oil embargo, showed 2.5 billion barrels of oil and 20 billion cubic feet of natural gas in the arid and semiarid areas of the West (Miller 1975). This is a combined energy equivalent of 6.5 billion barrels of oil. As of December 31, 1974, 61 percent of the total cumulative production and

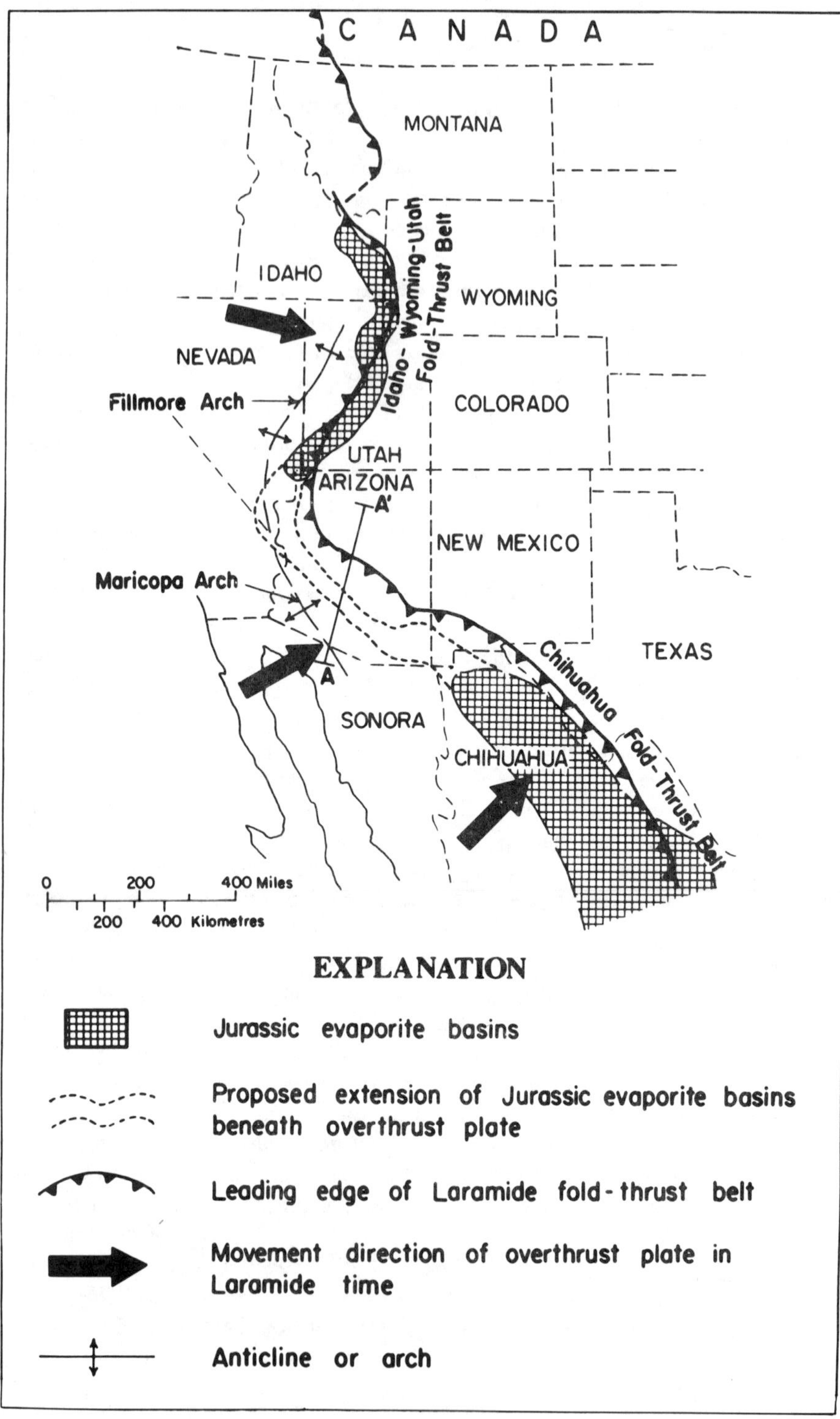

Figure 1.1 Projection of overthrust belt through Mexico and Western United States (Keith, 1979, State of Arizona, Bureau of Geology and Mineral Technology)

demonstrated reserves of U.S. petroleum were located in the arid and semiarid areas of the Western Rocky Mountain, the Northern Rocky Mountain, the West Texas and Eastern New Mexico, and the Western Gulf Basin regions (Miller 1975). These regions also contain an estimated 41 percent of the total U.S. undiscovered recoverable petroleum resources. Approximately 7 percent of these on-shore resources are on federal lands.

Tar sands. Tar sands are a much more limited energy resource than any of those previously mentioned. Occurring predominantly in Utah, these deposits are estimated to contain about 17 billion barrels of oil of which only about 10 billion barrels would be recoverable.[9] Over 85 percent of all U.S. tar sands are located on federal land.[10]

Geothermal energy. No geothermal energy resources have yet been developed in the arid and semiarid areas of the United States; however, a 50 megawatt demonstration unit is planned for construction in the Jemez Mountains in New Mexico near Los Alamos. Exploration programs are underway in most of the Western States under Department of Energy sponsorship or under Department of the Interior-Bureau of Reclamation sponsorship—the former for energy purposes, the latter for pure water purposes. Geothermal energy resources in these areas have not yet been fully characterized, but estimates for the electric power potential by the New Mexico Energy Institute for the five states of Arizona, Colorado, Nevada, New Mexico, and Utah show the possibility of between 4000 megawatts and 20,000 megawatts.[11] This would be equivalent to an energy production rate derived from between 100 and 300 million barrels of oil per year. Department of Energy goals for the Rocky Mountain Range and Basin area are for the development of geothermal energy equivalent to 793 million barrels of oil per year. In these five states an aggregate of 52.6 percent of the land is controlled by the federal government indicating that a large portion of the potential geothermal energy resource of this region will be controlled by the federal government.

Hydroelectric. Further development of hydropower in the western states, particularly in the arid and semiarid regions, is extremely limited by the availability of appropriate dam sites. Undoubtedly pumped storage sites will be selected for specific energy *storage* projects but pumped storage is not a primary source of energy and will not contribute to our overall energy economy.

PROBLEMS AND ISSUES OF ARID LANDS ENERGY RESOURCE DEVELOPMENT

As later chapters will elaborate on in more detail, the development of the energy resources of the U.S. West and Southwest will encounter a number of very difficult problems and issues. The lack of adequate amounts of water is one very major problem as is the environmental protection problem. Brown and Kneese (1978), publishing in *The American Economic Review,* listed three key issues which concern the Southwest as it faces the development of its huge energy resource:

1. *The environmental preservation issue.* This is an area of high aesthetic value with many natural parks, monuments, forests, wilderness and other federal and state reserves, an area of spectacular mountains, fertile valleys and sweeping plains.

2. *The resource revenues issue.* Who owns the energy resource located on the predominantly public lands of the western states? The region in question contains many of the nation's poorest sub-societies. An aggregate indication of this is found in the 1976 per capita income figures for Utah and New Mexico at 84 and 83 percent of the national level, respectively. More dramatic, however, are the economic conditions of the Indian tribes which inhabit the region. The 1970 median family income for the 125,000 member Navajo tribe of New Mexico, Arizona, and Utah was but 32 percent of the national level. How should revenues from energy resource development on these lands be distributed? The respective states have different philosophies about energy resource extraction which range from little or no taxation in Utah to what some consider as excessive taxation in Montana. Thus far, the states involved have demonstrated little or no attention to the use of energy resource tax revenues to improve the economic lot of the regions involved. The New Mexico State legislature, for example in 1977 rebated $100 million to New Mexico taxpayers rather than invest it developmentally within the state.

3. *The water issue.* The water issue is actually a whole set of issues which extend beyond the mere fact that water is a scarce commodity in this region:

    a. *the equity issue,* the ownership of water rights in many of the states concerned remains highly unsettled. While most western states have practiced a full appropriation of ownership rights, the federal government recently has promoted two other categories of rights—Indian and federal reserved—which in many arid areas in particular are in strong conflict with rights established under the prior appropriation system;

b. *the efficiency issue,* the institutional procedures of transferring water rights from one owner to another is a cumbersome process in many western states. Arizona, for example, with a farmer-dominated water law requires that water rights be transferred with land ownership and prohibits the separation of water rights and the land. Other states are more flexible in this issue;

c. *the environmental quality issue,* like air quality, water purity is an increasingly vital issue in arid and semiarid areas. This problem concerns the extension of the definition of legally accepted uses of water to include maintenance of stream flow as an aquatic habitat and for aesthetic reasons;

d. *the water development issue,* dramatically focused by President Carter's 1977 budgeting actions, assured that water development projects in the West will receive close scrutiny.

Throughout this chapter I have cited an estimation of the amount of energy resource which is on lands controlled by the federal government. This control intensifies the difficulty of developing these energy resources immensely inasmuch as the federal bureaucracy must be alert and sensitive to the full range of problems and issues engulfing energy resource development on these lands. Coal is a good example of the case in point. In spite of the severity of the energy shortage, not one coal land lease of any consequence has been issued since 1971! Efforts by the Department of the Interior to lease such lands in 1974 brought about a suit against the Secretary by the Sierra Club[12] and later by the Natural Resources Defense Council[13] which thus far has completely stymied increased coal resource development on these lands. The Bureau of Land Management estimates that it will not be until 1982 that coal land leasing of any consequence will be reestablished.[14] This is despite the fact that these lands contain the majority of the nation's low sulfur coal reserve and that the President has said that U.S. coal production must double by 1985![15] Further, an estimated three-quarters of the federal lands involved have been withdrawn from any form of mineral development *regardless of the need of the nation for these minerals* (Bennethum and Lee 1975). The importance of this statement is demonstrated by the fact that the Department of the Interior estimates that an average of 47 percent (27 percent in Idaho, 35 percent in Utah, and 61 percent in Wyoming) of the Overthrust Belt region has been withdrawn or severely restricted from oil and gas leasing in spite of the acknowledged fact that this area could be one of the most significant oil and gas provinces in the United States![16] Recently President Carter asked that 9.9 million acres in Arizona and the other contiguous states be withdrawn from use under the RARE II program.[17] Included in this request is the area which is a prime target area for Arizona's first geothermal energy demonstration program! These withdrawals have been made in the name of preservation and usage of the land for higher order uses than mineral and energy development.

Thus, while the U.S. arid region contains energy resources which are several times in excess of that of another arid region, the Middle East, it appears that there are many technological, sociological, and political problems to overcome before these resources can be put to use to solve our present energy dilemma.

NOTES

1. Energy from the West: A Progress Report of a Technology Assessment of Western Energy Resource Development, United States Environmental Protection Agency, EPA-600/7-77-072a, July 1977, p. 6.

2. Estimated at 450 billion barrels of oil remaining proven reserves. Energy: Global Prospects 1985-2000, report of the Workshop on Alternative Energy Strategies, MIT, McGraw Hill Book Company, New York, 1977.

3. Coal Resources of the United States, January 1, 1974, United States Department of the Interior, Geological Survey, Bulletin 1412 (1975), p. 35; also The Reserve Base of U.S. Coals by Sulfur Control, 2. The Western States, United States Department of the Interior, Bureau of Mines, Information Circular 8695 (1975), p. 7.

4. National Energy Policy: A Continuing Policy, Council on Energy Resources, The University of Texas at Austin, January 1978, pp. 313-381.

5. Energy from the West: A Progress Report of a Technology Assessment of Western Energy Resource Development, United States Environmental Protection Agency, EPA-600/7-77-072a, July 1977, p. 7.

6. News Release, Department of Energy, Grand Junction Office, April 25, 1979.

7. Ibid.

8. Op. cit. National Energy Policy: A Continuing Policy.

9. Final Enviromental Impact Statement = Proposed Federal Coal Leasing Program, United States Department of the Interior, Bureau of Mines, May 1974, pp. 8-21.

10. Op. cit. National Energy Policy: A Continuing Policy, p. 321.

11. Regional Operations Research Program for Development of Geothermal Energy in the Southwestern United States, Final Technical Report, June 1977 to August 1978, New Mexico Energy Institute, Las Cruces, New Mexico, January 1979, p. 84.

12. *Sierra Club* v. *Kleppe* originally *Sierra Club* v. *Morton*, 427 U.S. 396 (1976).

13. *NRDC* v. *Hughes*, United States District Court for the District of Columbia, Civil Action 75-1749.

14. United States Department of the Interior, Budget Request FY 1980.

15. The National Energy Plan, Executive Office of the President, U.S. Printing Office No. 040-000-0380-1, April 24, 1977, p. 96.

16. Final Report of the Task Force on the Availability of Federally Owned Mineral Lands, United States Department of the Interior, U.S. Government Printing Office No. 024-000-00834-7, July 1978, p. 56.

17. Press Release, Office of the White House Press Secretary, April 16, 1979.

REFERENCES

Bennethum, G. and L. C. Lee. 1975. Is our Account Overdrawn?, *Mining Congress J.* 61 (9):33-49.

Brown, L. and A. B. Kneese. 1978. The Southwest: A Region Under Stress, *The American Economic Review,* 68 (2).

Keith, S. B. 1979. The Great Southwestern Arizona Overthrust Oil and Gas Play, *Fieldnotes,* State of Arizona, Bureau of Geology and Mineral Technology, 9 (1):10.

Miller, B. M., et al. 1975. Geological Estimates of Undiscovered Recoverable Oil and Gas Resources in the United States, United States Department of the Interior, Geological Survey, Circular 725, pp. 4, 28, 29.

2

SOCIOECONOMIC AND CULTURAL IMPACTS OF ENERGY RESOURCE DEVELOPMENT ON INDIAN LANDS

Thomas Leubben

Leubben, Hughes and Kelly Associates
Albuquerque, New Mexico

SUMMARY

Being Indian is a philosophy and an ideology, not simply a racial characteristic combined with a culture of "primitive" ways. The purpose of tribal life is to perpetuate itself and to maintain eternally the cycles of physical and spiritual existence. Large-scale resource development forces fundamental changes in this native American ideology. The traditional economy and lifestyle is destroyed, individual moves into the wage economy will develop unconditional dependence on modern goods and services. Urbanization destroys the old ways, the old attitudes, and old relationships. The Indian cultural value system characterized by strong sense of place, strong family and clan relationship, strong sense of spirituality, lack of materialism, egalitarianism, and high regard for older members of the society is replaced by materialism, less sense of community and extended family, emphasis on education, emphasis on youth, devaluation of the position of older people, high regard for change and individualism. Decision making necessitated by resource development creates unreconcilable internal tribal conflict. Colonial economics are created which impact adversely on the existence and exercise of tribal sovereignty. Large influxes of non-Indians also impact adversely on tribal sovereignty.

INTRODUCTION

Those engaged in seeking and assuring a continuing adequate supply of energy for the United States have recently realized that there are substantial energy reserves located on Indian reservation lands, particularly on the reservations of the Western United States. For instance, it is reported that fully 80 percent of the nation's uranium reserves, 33 percent of the nation's low-sulfur, strippable coal, and 3 to 10 percent of the Nation's recoverable oil and gas is located on Indian lands (Jorgensen 1978).

The larger reservations which contain these resources are primarily located in the northern Great Plains and the Rocky Mountain Region. These areas are arid or semiarid, and scarcity of water is one of the important determinative factors in the impact of energy resource development on native cultures and the land which they inhabit.

Energy resource development on Indian land represents a new phase in the on-going confrontation between European technoculture and North American native peoples. To a large extent, this new phase is an unexpected development. When the reservation system was being created in the late 19th and early 20th centuries, the federal government was relatively careful to locate the intended permanent homelands of western Indian tribes in the economically least attractive areas of the Western United States.

The majority of Americans will readily assume that by now the western Indian tribes are sufficiently assimilated into the mainstream of American lifestyle and culture that the development of energy resources on their reservation lands is not a particularly disruptive or traumatic event, or one which presents any unusual problems. In fact, the interface between European and Native American culture remains quite unresolved. Indian tribes are still struggling vigorously, and with a unique commitment, to defend themselves against colonization and assimilation by Europeans. The initiation of large-scale energy projects on Indians lands has greatly increased the tempo of that struggle. The conflict rarely expresses itself with military violence, however, as in the past. Instead, the struggle is carried on with words, conferences, lawyers, courts, and occasional, usually nonviolent, demonstrations and confrontations.

Energy development on Indian lands will bring about some of the greatest changes the tribes have experienced since encountering Europeans. If trends continue as in the past decade, the toll on Indian people could be disastrous. It is important to understand the nature and significance of this process. It is representative of the confrontation between technological and non-technological cultures the world over, and it tells non-Indians much about themselves.

SOME KEY ASPECTS OF NATIVE AMERICAN CULTURE

There are two topics fundamental to understanding the impact of resource development on Native Americans. They are (1) the concept and practice of tribal sovereignty and (2) the nature of tribal life.

For Indian people, one of the most important aspects of tribal existence is tribal sovereignty. Tribal sovereignty is a sense of nationhood; it is the actual political and jurisdictional authority which tribes have exercised since time immemorial, and which they still retain in substantial measure. Without tribal sovereignty, Native American culture in any true sense cannot survive. Sovereignty is an indispensable part of the glue which holds a tribe together as a political and cultural entity.

There are very significant basic differences between the nature of tribal life and the nature of modern European technoculture. Simply stated, the purpose of tribal life is to perpetuate itself. Tribal life is not directed toward objectives, such as invention and innovation, economic progress, or the solution of the problems of all mankind. Tribal life is cyclical and circular, rather than goal-oriented and linear. Much of the energy in tribal culture is directed to maintaining the eternal cycles of physical and spiritual life. It is also extremely important to perpetuate tribal relations, clan relations and strong family relationships.

By contrast, much of the energy of European society is directed toward economic development and capital accumulation. Many of the older values which Europeans shared with tribal peoples as a result of European tribal origins have been sacrificed for the attainment of economic objectives. This process has replaced a tribal value system with a set of values more convenient and efficient for a technological culture.

Europeans habitually underestimate the strength and integrity of tribal cultures. For a great proportion of Native Americans, being Indian is not a racial issue or a problem of poor education and a low standard of living so much as it is a political and cultural ideology. Native Americans have historically believed, and the majority still believe, that Native American lifeways are superior to the European cultural system, or are at least different in such important qualitative ways that an attempt to reject European attitudes and lifestyle is justified.

Traditional Native American value systems have the following general characteristics:

1. Family identity and family ties are extremely strong. Extended family groups living in close proximity to one another are the norm. Clan, sacred society, and tribal relationships are very important and are maintained.

2. The people possess a strong sense of geographical place. Every rock and hill has a name and a special significance. The area traditionally

inhabited by a tribe provides that tribe and its members with a distinctive sense of identity and a distinctive role in the cosmic dance. The area within the traditional tribal boundary is a cosmos in itself; it is *the* tribal world and most of the history and mythology of the tribe is located geographically within it.

3. Land is not a commodity as in the white man's world. The land within the aboriginal tribal boundary is not owned by anyone; it is used and occupied by the tribe under a sacred mandate. The land must be respected and preserved. The tribe would no more voluntarily sell or abandon its homeland than mankind would voluntarily abandon the planet.

In May 1971, when the Hopi traditionals filed suit against the Peabody Coal Company and the Secretary of the Interior seeking to cancel the Hopi Black Mesa Coal Lease, the elders attached a statement to the complaint which states this attitude well:

> Hopi land is held in trust in a spiritual way for the Great Spirit Massau-u. . . . The area we call "Tukunavi" which includes Black Mesa is part of the heart of our Mother Earth. Within this heart, the Hopi has left his seal by leaving religious items and clan markings and plantings and ancient burial grounds as his landmarks and shrines. . . .
>
> This land was granted to the Hopi by a power greater than man can explain. Title is vested in the whole make-up of Hopi life . . . The land is sacred and if the land is abused, the sacredness of Hopi life will disappear and all other life as well.[1]

Private ownership of land has always been foreign to Native American cultures. For the Hopis, land cannot be owned because it belongs to the spirits K'taimetah and Massau-u. Those who use the land must perform ceremonies to take care of it, and the land cannot be bought or sold (Clemmer 1978).

For the Navajo, land may be important in at least five primary ways; (1) as a locus of sentiment, (2) as the locus of cognition and memory, (3) as the locus of religious ceremony and belief, (4) as a provider of economic subsistence, and (5) as the locus of family and clan activity and relationships (Schoepfle, 1978).

4. Tribes are by nature spiritual communities and Native American people have a pervasive sense of spirituality and commitment to religious practices. For the more traditional people, maintaining spiritual and religious commitments is far more important than any economic objective. This is consistent with a notable lack of materialism. For instance, the Navajos do not raise sheep as a business. A herd of sheep does provide a basic subsistence, but spiritual motivations and family commitments are

also important reasons for maintaining the herd (Hausenpflug 1979, personal communication).

5. Native American culture is also characterized by what has been called the cultural ideology of generosity (Lamphere 1979). Status and respect are obtained by giving away goods, rather than by accumulating goods as in modern non-Indian America (The Hau de no sau nee 1978). In Navajo culture, a person who attempts to accumulate goods is thought to be mean or stingy (Schoepfle 1978). The Navajo have developed a whole set of ideas which back up cooperation and redistribution of goods. Giving unstinting aid whenever it is requested or needed is extremely important (Lamphere 1979). There is continual gift-giving and borrowing within the extended family. There is also a wide circle of kin who depend on each other and who ask for help whenever it is needed and who give help whenever they can. By this method, goods and income are redistributed and people are protected against variations in income and economic bad times (Aberle 1969). This Native American ethic of generosity has actually been criticized by proponents of European culture because it inhibits the formation and accumulation of capital which is so essential to economic progress.

6. The process of redistribution of goods and income which results from the ethic of generosity fosters an economic classlessness and political egalitarianism. Chiefs and leaders do not hold a cultural or structural right to economic exaltation; they generally share the economic situation of their own extended family (The Hau de no sau nee 1978).

7. Another extremely important concept in Native American culture is the obligation of stewardship with respect to the land. The Council of the Six Nations (Iroquois Confederacy) is required by tradition to contemplate the well-being of the seventh generation in the future when making decisions. (The Hau de no sau nee 1978). As the Hopi have said, "this land is a sacred home of Hopi people . . . it was given to the Hopi people the task to guard this land . . . by obedience to our traditional and religious instruction, by being faithful to our Great Spirit Massau-u . . . we have never abandoned our sovereignty to any form, power or nation . . . ." (Clemmer 1978). Gouging the earth is inconsistent with Hopi ethics and religion (Clemmer 1978).

8. Maintenance of the balance and harmony of nature is also an extremely important objective. Hopi culture can be seen as "an ideally patterned relationship between human activity and environmental processes. The tranquillity of this relationship is achieved by balancing an intensity of purpose and behavior with an unfolding of life processes in a land of ephemeral resources" (Clemmer 1978). The purpose of tribal ceremonial life is to maintain the balance of nature and the harmonious relationships

of everything which plays a role in the natural world (personal communication, Joe Cook, Creek Medicine Man, 1976).

9. One of the most striking aspects of North American tribal culture is the requirement of consensus or unanimity in group decision-making. Everyone in the group, community or tribe is entitled to have his say about important issues which affect the group as a whole, and no decision or course of action can be adopted until everyone affected is in agreement. If a consensus cannot be reached, then it is not appropriate to take whatever action may be implied by the pending decision. The requirement of unanimity could hold up a proposed development project for years, or even a generation, but in the traditional value system it must be respected nonetheless.

10. Cultural and tribal survival over the long term is a transcendent objective. Anything which threatens or destroys the social and cultural cohesiveness which is required to maintain the tribal entity is extremely disturbing to traditional communities.

CHANGES EXPERIENCED AS A RESULT OF ENERGY RESOURCE DEVELOPMENT WHICH AFFECT NATIVE AMERICAN SOCIAL AND CULTURAL SYSTEMS

Economic modernization of tribes and reservations brings about a definite change in their value systems. This has usually been a relatively slow process, to the great frustration of the United States Government and the Department of the Interior. Large-scale energy resource development accelerates these cultural changes. The more that Indian people move into a modern American lifestyle, the more change takes place in their value systems. Cultural disruption and cultural changes parallel the changes in values.

Traditional peoples do not change their value systems by consciously and voluntarily embracing the new value of technological, industrial society. Rather, they experience a complex of disruptions of traditional culture flowing in from outside the reservation. The traditional culture, and the value system embraced by individuals, changes under the pressure of induced environmental and social disruptions.

Many of the economic and cultural changes brought about in the Southwest by energy resource development are obvious to anyone who drives through the region. Of course, many of these changes are part of a much larger and extremely complex historical event, which is the overall modernization and industrialization of reservation life.

The influx of federal money through health, education, housing, employment and many other programs has probably had a greater impact on reservation life than energy resource development. Nonetheless, resource development is an extremely significant part of the changes which are now occurring. From many viewpoints these changes can be judged as good and desirable. They include better housing, better health care and better educational opportunities. To some extent, energy resource development may contribute to these improvements, although that has been seriously questioned (Arthur 1978). It may be that the negative impacts of resource development will far outweigh any improvements which it may bring to reservation life. These negative impacts are the unavoidable result of the confrontation between essentially incompatible cultural systems, the tribal and the industrial. The negative impact of energy resource development on Indian lands is magnified by the fact that the development takes place over the course of a few months or a few years, rather than generational spans of time which would allow tribal culture to change gradually and perhaps adapt to the new order, instead of being shattered by it.

The new value system which industrialization pressures the tribal population to embrace is characterized by a greater materialism. Material accumulation and economic advancement require wage jobs and industrial discipline. This produces the cash required for modern goods and services. This is consistent with the American emphasis on individualism as a lifestyle. Some jobs are produced by the new developments, but the old economic classlessness and egalitarianism of traditional life suffer considerably.

A recent survey by the Shiprock Research Center of Shiprock, New Mexico confirms that energy development tends to erode the economic egalitarianism of traditional Indian cultures and develop economic class stratification. In the Four Corners area of Utah, New Mexico, Colorado and Arizona, which has been heavily impacted by major energy developments, the median Navajo household income has declined since 1970, even though the mean household income has increased (Arthur 1978). This increasingly unequal distribution of income erodes the social and political position of those at the lower income levels. This is the economic syndrome common to third world nations; the rich are getting richer and poor are getting poorer. Although the ethic of generosity and cooperation which developed in the Navajo subsistence economy still functions in a wage dependent economy to protect against low and fluctuating levels of income (Lamphere 1979), movement toward a less thorough redistribution of income and more material accumulation by certain families is unavoidable.

Energy resource development has definitely augmented the breakdown of the traditional extended family system and a movement toward the nuclear family organization characteristic of industrial societies. Studies of the impact of the construction of the Navajo generating station at Page, Arizona showed that Navajo people in the area tended to have smaller families and to be less involved in reciprocal relationships with kinsmen than was true of other areas of the reservation (Halloway, Levy and Henderson 1976). Largely because of the geographical relocations which are often necessary for wage workers, there is less sense of community and fewer extended families. Nuclear families, composed of father, mother and children, are desirable in industrial societies because, not being burdened by extended family ties, they are extremely mobile.

The research in the Page area showed that Navajo household heads are younger than those in areas not impacted by energy development. This is because the better educated and more skilled Navajos leave their home communities in order to obtain the higher paying jobs near the energy developments. The result is that other areas of the reservation are drained of younger people. These more rural areas come to be inhabited by residual populations comprised of the least skilled, the older and the more dependent people (Halloway, Levy and Henderson 1976).

It is obvious that Indians who move into an area to obtain industrial employment move away from the traditional lifestyle toward an American industrial lifestyle. These individuals must also develop industrial discipline; that is, they must commit themselves to show up at work on time, stay on the job, and come to work day after day. Industrial discipline conflicts with the traditional lifestyle which often demands attendance at religious and healing ceremonies which may last as much as four or sometimes eight days, without regard for construction or production schedules.

Along with industrial lifestyle goes the emphasis on youth which characterizes modern America. Younger people are capable of performing the labor and construction jobs which are created by energy resource development. They acquire an income vastly in excess of anything their elders ever imagined possible. As the younger people move ever more toward technoculture, the role of the older people in the Indian culture, and to some extent women as well, is devalued.[2] The effects of the social and cultural changes are seen particularly with the older people. As the native value system becomes more oriented toward economic success, and less concerned with the traditional values of harmony with nature and religious practices, the function and status of older people is undermined. The social and cultural changes also affect older people physically. Older people tend to die sooner because their lives lack the significance that they once had, and their value system may no longer be respected. This is particularly true when the land itself is destroyed by strip mining and the older people who were herders have nothing to do anymore.

For the Navajo, the large tribal land base has provided a shelter for people living a traditional lifestyle. The opening up of remote reservation areas, which usually occurs as a result of resource development, destroys this sheltering ability of the reservation and exposes them, their families and their clans to the impacts and the pressures of the industrial economy.

Strip mining violates the most fundamental tenets of Indian ethics and religion (Clemmer 1978). For traditional people who take seriously the concept of their stewardship of the land for future generations, energy development is anathema. Despite the contentions of the energy companies that the land can be rehabilitated, the traditional person knows that things will never be the same again, and there is at least a substantial risk that the land will not be effectively rehabilitated in the agricultural sense. In arid areas, traditional water supplies are also adversely affected. For the traditional Indian, to accept or support large-scale mining is to abandon most of the old teachings about land and its role in the sacred order of things. Most seriously, a felt loss of culture or cultural values is a felt loss of identity, of a part of one's very being.

Perhaps the most socially and culturally destructive impact of energy development on Indian communities is the bitter factionalism that it has generated in communities directly affected by development. Typically, the decision to lease tribal land for coal mines, uranium mines, power plant sites and other facilities has been made by the central tribal government without regard for the communities affected. When the Hopi Nation leased land for strip mining on Black Mesa, there were no open hearings, no community meetings, and no administrative disclosures by either the tribal government or the Bureau of Indian Affairs (Clemmer 1978). This fact created confusion, rumors, and discord among the affected communities. Even when some notice has been given to affected communities, the information provided about the project and its impacts has usually been woefully incomplete, and little attempt has been made to ensure that people actually understand what is proposed.

The Burnham Chapter of the Navajo Nation is a classic example of a community traumatized by the threat of energy development. In 1968, the Navajo tribal council in Window Rock agreed to lease 40,000 acres of land in the Burnham Chapter to the Consolidation Coal Company of the El Paso Natural Gas Company for strip-mining. Before strip-mining can begin, however, the energy companies must acquire the grazing rights of Navajos living in the lease area. Strongly traditional people in the Burnham Chapter have consistently opposed strip-mining and have initiated litigation intended to delay or stop it. Other people in the chapter have seen the mine development as an opportunity to sell their tribal grazing permits to the companies for relatively astronomical prices. The problem is eloquently stated by an elderly Burnham man (Hausenpflug 1979):

> Before the Council started no one owned land—we used it and our Mother the Earth gave us her gifts. We moved as we chose to move, as long as we did not bother any one else. Then the Council gave us grazing permits and the grazing committee gave us land which they told us was ours to keep and we should not move off it. This made us all jealous of each other, and my brothers and sisters southeast of here have sold their land to the coal company. People become so greedy for their money that they have forgotten that the land is their mother. My brothers and sisters have sold their land and now they are trying to bewitch me so that they can get the money for my land.[3]

While the breakdown of extended families into nuclear families seems to have been initiated by the creation of tribal grazing permits in 1941, the purchase of grazing permits by the energy companies has fostered this development (personal communication, Hausenpflug 1979).

The energy companies began buying grazing permits in late 1976. The bitter factionalism of the community, accompanied by increased accusations of the practice of witchcraft (an index of conflict within a Navajo community) coincided with the initiation of the grazing permit purchase (personal communication, Hausenpflug 1979). For the Navajo, the relationship between the people and the land is intricate and long-lasting. The nature and organization of the tribal land tenure system has a strong influence on cultural practices. The evolution of social structures is intimately related to the evolution of land use rights and tenure. Alteration of the land tenure system tends to produce corresponding cultural changes (personal communication, Hausenpflug 1979). At Burnham, people tend to view the selling of grazing permits as equivalent to selling the land. Whereas before, no one owned land, it is now being sold for high prices. This fact can be expected to produce a significant restructuring of attitudes and cultural practices.

The breakdown of the Burnham Navajo Chapter organization and the developing factionalism has caused people not to be concerned with the plight of their neighbors (Hausenpflug 1979). Even though actual mining operations have not yet begun, the anticipated development has already resulted in decreased community and kin cooperation.

As might be expected, when energy development requires relocation of people occupying the land, social and cultural impacts are especially severe. Navajos who have opposed relocation due to energy development have expressed three basic reasons for their opposition; (1) they expect the disruption of their normal kinlife, (2) they lose their sheep and the use of the land and (3) they anticipate economic hardships as a result of the loss of the sheep and the land and relocation (Schoepfle 1978).

Relocation splits up the kin groupings. This splits up economic units based on kin groupings, and is extremely upsetting to people living in a system based on close kin relationships. Possession of a sheep permit and the ownership of a herd enable people to help one another out in the traditional ways, which is extremely important in the Navajo value system. Schoepfle has also noted that relocation produces psychosomatic illnesses.

The rehabilitation of strip-mined land results in a regular rolling terrain radically different from the original flat and broken terrain. Traditional people may use geographical features as mnemonic aides for remembering traditional lore and knowledge and the procedures for implementing and teaching it. The natural features of the land surface may provide a kind of cognitive map. This map is obviously destroyed by strip mining, along with the knowledge and procedures that it incorporated. In addition, many Navajos attribute great importance to certain sacred stones and sacred areas which are destroyed by mining (Schoepfle 1978).

Obviously, the loss of a Navajo grazing permit through its sale by an individual, or its sale by a person from whom the individual might inherit, forces that individual out of the traditional cultural pattern and into an industrial or urbanized lifestyle. This is true because there is usually no place left for a displaced person to go and reestablish the traditional lifestyle based on sheepherding.

Still another effect of the industrialization of rural Native Americans is a decrease in their commitment to the tribe. Navajo workers on the Page power plant and Black Mesa mining projects exhibited few allegiances to the tribe and were largely people who had gained skills off the reservation. They returned to the reservation to enjoy the benefits of the construction work, but planned to return to their previous off-reservation residences when the work ended (Robbins 1978).

SOME PARTICULAR PROBLEMS ASSOCIATED WITH ENERGY DEVELOPMENT

Impairment of Tribal Sovereignty

Large scale resource development threatens tribal sovereignty in a variety of ways. Development of large projects brings an influx of non-Indian construction workers and entrepreneurs of various types. The sparsely populated, arid areas of the Western States where large energy resource recovery projects are being planned are usually unable to supply project labor demands. A large pool of unskilled labor from other regions of the country must necessarily migrate to the projects (Little 1978). Although this problem is greatest during the construction phase of the project, many non-Indians will stay in the area when construction is completed. Also, the secondary economy which a project creates invariably brings in additional numbers of non-Indians.

To the extent that the increased non-Indian population is located on the reservation, it can create serious problems. Recently, the United States Supreme Court decided in the case of *Oliphant* v. *Suquamish Indian Tribe*[4] that Indian tribes do not have criminal jurisdiction over non-Indians on the reservation. If Indian tribes cannot control the criminal behavior of non-Indians on the reservation, some other governmental entity must. The state within which a given reservation lies is the most likely entity to fill the gap. This incursion into the tribal jurisdictional domain represents a substantial loss of tribal sovereignty. If tribes themselves are ultimately unable to control non-Indians on the reservation, the adverse political impact of substantial increases in non-Indian population will be great.

On the Crow Reservation, tribal members have expressd concern about the non-Indian population increase resulting from energy developments. They have indicated a fear that Indians will become a minority on their reservation, that their culture will die out, and that they will lose control over the land, and ultimately their lives (Owens 1978). This fear is very real. Today, over 60 percent of all Indian reservations have a non-Indian majority population living on the reservation. In every instance this has led to serious problems in the exercise of tribal sovereignty and tribal jurisdiction. Typically, if they are subject to tribal government, the non-Indian majority on an Indian reservation will eventually demand political rights in the decision-making process and the process of government. History shows that once the non-Indian population has an upper hand, they will not respect the legal and social history of the tribe which is the basis for tribal sovereignty and jurisdiction.

It is a universal rule that Indian and tribal influence in a region declines as the non-Indian population increases. A large increase in non-Indian population adjacent to the reservation generally has an adverse effect on tribal options. There have been intense struggles for control of local school boards and school systems, and for control of county government.

Studies of boom communities in the western United States have recognized that indigenous political structures usually undergo radical changes as a result of the influx of large numbers of outsiders. Political control, as exhibited by leadership in the community, is likely to shift to newcomers (Little 1978). For instance, in eastern Montana, the political power base is shifting from the long established ranchers to the newly arrived industrial personnel (Little 1978).

The Boomtown Syndrome in the Tribal Context

The boomtown syndrome has received considerable study in recent years. A boomtown results from large-scale, sudden, economic development which creates an extrordinary population increase in a short period of time in a smaller community. The severity of social impacts are directly related to the intensity and rapidity of population growth. If growth is too rapid, communities are unable to meet the social and service demands of the populace and a general social and political breakdown occurs (Little 1978). High rates of divorce, depression, alcoholism, and suicide result (Little 1978). These problems increase more rapidly than the rate of population increase (Little 1978). Crime rates increase and intracommunity conflicts increase. The high crime rates in boomtowns signify a serious deterioration in the quality of life (Little 1978). The high crime rates appear to be due to the extremely transient population. Construction workers, particularly, do not identify with the town or its institutions and organizations because their residency is temporary.

The Hopi, Navajo, Northern Cheyenne and Crow Indian communities have all been exposed to some of the boomtown impacts. The social problems which Little has documented in non-Indian boom communities are compounded when the interface between technoculture and tribalism is involved. Whereas newcomers to a non-Indian boom community are integrated into the community structure only with difficulty, they can be integrated almost not at all in an Indian community. Non-Indian newcomers do not possess the cultural background necessary to participate in the Indian community. They do not possess the tribal membership which is necessary to participate politically in the Indian community. Moreover, they typically have an attitude about Native Americans which is extremely conducive to racial conflict. Little found that local Anglo populations "generally view Navajos, as well as other tribal groups, as living in a separate world, inept and undereducated if not indolent, and cared for by misguided bureaucrats and government largesse" (Little 1978). In Page, Arizona, and Hardin, Montana, Little found a general attitude that exploitation of the local Indian populations was acceptable. Inconsiderate and even violent treatment of Native Americans is common (Little 1978).

Little's findings confirm the operation of a larger, complex sociological process. Perhaps the most tragic and disturbing effect of the industrialization of non-technological cultures is that the native population inevitably becomes the lowest class in the economically and socially stratified industrial society. Although they move out of the traditional culture toward technoculture, giving up traditional values by degrees and gradually embracing the value system of industrial society, they will not be fully integrated into the new culture for many generations, if at all. Racial and cultural prejudice by the dominant society, lack of facility with the foreign culture and language, and continuing identification with aspects of the old culture prevent complete and effective assimilation into the new culture. Tribal people in transition, having given up significant portions of traditional culture, often find themselves without a firm tribal Indian identity, yet not really a part of the dominant non-Indian industrial society. They have given up a vital part of themselves, but have not gained a compensating position or identity in the new order. They are culturally disenfranchised. Throughout the world, such people occupy the bottom rung of the social and economic structure of industrial society, and become an exploited and abused racial minority.

If the residents of a new town with no history, such as Page, Arizona, and newcomers to boomtowns, are not strongly committed to their local communities, the commitment of such persons to Indian communities is nil. Construction and energy workers are intimately tied to a belief system that prioritizes progress and material productivity; they generally view the ancient, economically inefficient, non-technological tribal lifeways with scorn. Indian social control mechanisms cannot affect their behavior. Formal tribal law enforcement mechanisms are ineffective because the tribe has no juridiction off reservation and no criminal jurisdiction over non-Indians on the reservation.

In off-reservation border towns, disregard for and violation of Indian civil rights is common in normal times. Under boom conditions the problems escalate. A boomtown is not a healthy place for an Indian. Stereotyped attitudes towards Indians prevail, and there is no commitment to tribal sovereignty or Native American value systems.

Effect of Energy Resource Development on the Tribal Political Process

One of the most serious of the social and cultural impacts of resource development on Indian lands is its effect on tribal cohesiveness and the tribal political process. The decision to lease tribal lands for coal, uranium power plant sites and other facilities typically produces a bitter political struggle within the tribe. This has been dramatically evident in the Navajo and Hopi tribes when leases were signed for major resource development projects. In almost every instance, the traditional people have vigorously opposed the projects.

In a tribal context, the significance of this kind of event is much different and much more serious than would be the case with an analogous situation in a non-Indian community. In a non-Indian community, the formation of groups to advocate or oppose a given action is simply part of the legitimate political process. In one way or another, the issues are presented to the voters and decided by majority rule. In a tribe, however, an event of this kind represents a major cultural breakdown, and fosters the disintegration of tribal relations. The European principle of majority rule is contrary to Indian culture and values. The traditional Indian view is that important decisions must be essentially unanimous among those enfranchised to participate in the particular decision-making process. The tribal way of proceeding is by consensus. As long as anyone objects to a proposed action, a decision cannot and should not be made. A decision made by outvoting the opposition violates their rights and guarantees the existence of a dissatisfied and dissident minority.[5]

Traditional tribal governments are very poorly designed to make decisions in the volume and with the speed required in a modern industrial society. Industrial development decisions usually cannot wait for the months, or sometimes even the years which it may take for a consensus to develop in an Indian community. In many tribes, such as the Navajo and Hopi, there never was a centralized traditional tribal government which could make binding commitments on behalf of the entire tribe, or lease or dispose of land. Land was not owned by anyone, and no one had the right to dispose of it.

During the 19th century the Navajo tribe consisted of a number of bands under the cooperating but separate leadership of local headmen. The Hopi Nation was a culturally affiliated group of independent city states, similar to ancient Greece. The present tribal governments of both of these tribes were designed and imposed by the Department of the Interior to serve the objectives of the federal government. A principle objective was the creation of an authority which could negotiate and execute agreements on behalf of the tribe as a whole. For instance, the United States Department of the Interior initiated efforts to organize a centralized Navajo tribal council as a result of the discovery of oil at Hogback, New Mexico, on the Navajo reservation, in 1922. The council was needed to represent the entire reservation on matters of land leasing and negotiations (Hausenpflug 1979). The first Navajo tribal council was organized by the Department of the Interior, and its form was dictated by Interior Department regulations (Hausenpflug 1979).

In the late 1920's, the Bureau of Indian Affairs began trying to organize Navajos into local chapters. The chapters were necessary to provide a political link between individual Navajos and the centralized tribal council (Hausenpflug 1979). This was part of the process of building a centralized tribal government with which the Department of Interior and mineral exploitors could deal.

In creating the Navajo chapter system, and the centralized governments of many tribes, the Bureau of Indian Affairs imposed majority-rule voting in the place of the traditional rule of consensus. The traditional people have never accepted the change from the highly individualistic, independent, directly representative and consensus-oriented traditional political process to a centralized tribal government. They often refuse to participate in federally-sponsored tribal governments. The approval and execution of a mineral resource lease by such a centralized tribal government without the participation of the traditional decisionmakers in the tribe is offensive and unacceptable to the traditional people. They do not feel a part of such a tribal government or bound by its decisions.

In the Burnham Chapter of the Navajo tribe, some Navajos have stated that because consensus based decision-making was bypassed by both the centralized Navajo tribal council and the mining companies, hostile factions formed. The factions formed because the mining issue was not kept on the tribal agenda for discussion until all people felt satisfied and until a decision was unanimous. The formation of these factions was reinforced by a majority vote decision forced on the chapter by the tribal government (Schoepfle 1978). Those who were outvoted are still vigorously fighting the proposed mining project.

The loss of consensus decision-making has been extremely disturbing to the members of the Burnham Chapter. They are very anxious to restore the consensus decision-making process. There is, however, little hope of doing so in view of the overwhelming political and economic forces bearing on the community.

When the rule of consensus is violated in tribal decision-making, a festering political problem is produced which often explodes into physical occupation of the project site, or even physical violence. The opposition of dissident tribal factions can be a major problem for resource companies, and more than one project has had to be scrapped because of such opposition.

Non-traditional tribal governments are virtually always dominated by the "progressive" elements in the tribe; the people who accept the authority and interference of the Department of Interior and who do not object to doing things which offend traditional values. This is because the Interior Department dominated tribal governments receive the legal and financial support of the federal government. In 1955, the Interior Department pushed aside the traditional Hopi government in favor of the more tractable Hopi tribal council. One Hopi noted that the tribal council had always been a "government council," not a Hopi council (Clemmer 1978).

The nature of the tribal form of social organization seems to require that political, religious and social functions be unified. Modern tribal governments, often modeled after the federal government, separate political from social and religious processes. Typically, in non-traditional, constitutional tribal governments there is an implicit recognition and application of the American political doctrines of separation of church and state and separation of powers. This separation of the political and religious functions sets into motion forces which tend to destroy tribal culture. Since the political processes are dominated by the more progressive elements, traditional religion and culture become irrelevant in the process of making the tribal decisions which really count. Decisions are often made which impact severely on traditional values and traditional practices. The result is the cultural disintegration of the tribe.

Under these circumstances the traditional people will form a permanent opposition block and engage in a long term struggle to regain control of tribal government and save the culture. The progressive elements, with the support of the Department of the Interior and a host of outside interests, always win this battle.

CREATION OF A COLONIAL RESERVATION ECONOMY

Many observers of the changes in reservation life as a result of resource development have expressed concern that large-scale project development typically creates a colonial economy on the reservation. A colonial economy can be characterized by the following:

1. Most development capital comes in from outside the area.

2. Control of economic enterprises remains outside the area; unseen and unreachable.

3. The native population participates as a labor pool and not as owners or managers.

4. Management and investment policies are controlled from outside and profits flow to the outside.

The development of a colonial economy is a serious setback to tribal sovereignty and self-determination. The forces at work in colonial economies and boom communities always conflict with native values and lifestyles. For the reasons reviewed above, traditional Native American culture stands in the way of the all-out economic development usually sought by non-Indian interests. In the 19th century, this conflict resulted in the anti-Indian policies of land allotment and cultural assimilation. Today, the same forces are directed against tribal sovereignty and jurisdiction.

Thus far, energy resource development on Indian reservations has been conducted almost entirely with the pattern of colonial economic development. Underdeveloped nations around the world have resoundingly rejected this pattern, and American Indian tribes are struggling to find alternatives.

THE FAILURE OF THE PROMISE OF ECONOMIC DEVELOPMENT

Unfortunately, many of the loudly proclaimed benefits to tribes from large-scale energy resource development have failed to materialize. Indian employment, almost always the most attractive aspect of energy development, has been disappointing at best. In 1977, there were only two Hopi employed in the entire work force at the strip mines on Black Mesa (Clemmer 1978). In 1978, only 2.7 percent of the entire Navajo work force was employed in energy industries. Wages paid to Navajos by the energy industries in 1974 added only $109.00 per capita to annual reservation incomes (Robbins 1978). The economic well-being of Navajos in the Four Corners Region has not improved significantly despite extensive development of coal strip-mines and power plants (Arthur 1978). In the Black Mesa area, increased job opportunities have only resulted in a lower level of unemployment and slightly higher than average Indian annual incomes. A large proportion of households still remain on welfare (Halloway, Levy and Henderson 1976).

Whether further energy development on the Navajo reservation will significantly assist the tribe to achieve economic self-sufficiency and a materially higher standard of living is open to serious question. Robbins has concluded that "the Navajo Nation goal of a viable self-sufficient economy cannot be achieved by the energy industries as they now exist and function in the Navajo economy. In fact, the Navajo economy, as measured by personal and tribal income, is steadily losing ground each year in comparison with that of the United States" (Robbins 1978).

The extractive industries are characteristically capital intensive. Since the processing and use of the mineral resources mined on Indian lands invariably takes place off reservation, there is little or no secondary impact through the creation of spin-off industries in the tribal economy.

CONCLUSIONS

Large-scale energy resource development on Indian lands produces a host of impacts which ultimately restructure Indian life and culture. Industrial societies world-wide tend toward a cultural norm which I have called "technoculture." Energy resource development on Indian lands greatly accelerates the process of tribal people moving into technoculture. It also impacts in major and specific ways which are negative by almost anyone's judgment. For a great many Indians, the process is involuntary; they do not want to suffer the cultural and lifestyle losses which large-scale energy development produces. Increasing consciousness in the Indian community and greater access to information has resulted in native opposition to almost every major development proposed or attempted in recent years. Residents of the Dalton Pass area of New Mexico, for example, passed a resolution protesting the adverse impacts of uranium mining; including the danger to the health and well-being of people and livestock from air, water and land contamination; the influx of outsiders; the stress on local services and facilities; disrespect toward Navajos and Navajo culture by non-Indians and destruction of Navajo sacred sites. Doubt about the durability and significance of benefits from energy development has become widespread in the national Indian community.

The impacts of development on Indian lands are important and relevant to the larger society, and should make a difference to some people. Those who are involved in energy development on Indian lands should be aware of the real impacts. It is not simply a question of providing energy, making a profit and bringing unmitigated economic benefits to an underdeveloped part of the United States. The human impacts of resource development on reservations are not merely incidental and academic changes to an outdated lifestyle. Something extraordinary and irreplaceable is being destroyed; something which industrialized technoculture no longer understands.

NOTES

1 Plaintiff's brief, *Starlie Lomayaktewa, et al.,* v. *Rogers C. B. Morton and Peabody Coal Company* (D.D.C. May 14, 1979), Exhibit A, p. 1.

2. Little (1978) noted that the kind of economic growth that characterizes boomtowns causes increased social stratification in a community, and a loss of status for women and the elderly.

3. *National Indian Youth Council* v. *Andrus,* Civil No. 78-586-C, United States District Court for the District of New Mexico.

4. 98 S. Ct. 1011 (1978).

5. The consensus rule may be a necessary principle of pretechnological tribal political life where the development of a strongly dissatisfied minority could lead to the breakdown of the tribal political process and the dispersion or splitting up of the tribe.

REFERENCES

Aberle, D. F. 1969. "A Plan for Navajo Economic Development." *Development Prospects and Problems,* Vol. I, part 1 of *Toward Economic Development for Native American Communities.* Washington, D.C.: U.S. 91st Congress, 1st session, Joint Economic Committee, Subcommittee on Economy in Government, U.S. Government Printing Office.

Arthur, H. 1978. "Preface." *Native Americans and Energy Development.* Anthropology Resource Center, Cambridge, Massachusetts.

Clemmer, R. O. 1978. "Black Mesa and the Hopi." *Native Americans and Energy Development.* Anthropology Resource Center, Cambridge, Massachusetts.

Halloway, D. G., J. E. Levy, and E. B. Henderson. 1976. "The Effects of Power Production and Strip Mining on Local Navajo Populations." *Lake Powell Research Project Bulletin.* No. 22. Department of Anthropology, University of Arizona, Tucson, Arizona.

Hasenpflug, J. F. 1979. "A History of Navajo Land Tenure and Land Use Rights." (Unpublished manuscript).

Jorgensen, J. G. 1978. "A Century of Political Economic Effects on American Indian Society: 1880-1980." *Journal of Ethnic Studies,* 6.

Jorgensen, J. G., S. H. Davis and R. O. Mathews. 1978. *Native Americans and Energy Development.* Anthropology Resource Center, Cambridge, Massachusetts.

Lamphere, L. 1979. "Traditional Pastoral Economy." *Economic Development in American Indian Reservations.* Native American Studies, Development Series No. 1, University of New Mexico, Albuquerque, New Mexico.

Little, R. L. 1977. "Some Social Consequences of Boomtowns." *North Dakota Law Review.* University of North Dakota School of Law, 53, Grand Forks, North Dakota.

Little, R. L. 1978. "Energy Boom Towns: Views From Within." *Native Americans and Energy Development.* Anthropology Resource Center, Cambridge, Massachusetts.

Owens, N. J. 1978. "Can Tribes Control Energy Development?" *Native Americans and Energy Development.* Anthropology Resource Center, Cambridge, Massachusetts.

Robbins, L.A. 1978. "Energy Developments and the Navajo Nation." *Native Americans and Energy Development.* Anthropology Resource Center, Cambridge, Massachusetts.

Robbins, L.A. 1979. "Structural Changes in Navajo Government Related to Development." *Economic Development in American Indian Reservations.* Native American Studies, Development Series No. 1. University of New Mexico, Albuquerque, New Mexico.

Schoepfle, G. M., K. Y. Begishe, R. T. Morgan, P. Reno and J. Begay. 1978. "A Study of Navajo Perception of the Impacts of Environmental Changes Relating to Energy Resource Development." Navajo Community College, Shiprock Branch, Shiprock, New Mexico.

The Hau de no sau nee. 1978. *A Basic Call to Consciousness.* Awkesasne Notes, Mohawk Nation, Via Rooseveltown, New York.

3

ARID LAND INHABITANT ATTITUDES TOWARD ENERGY RESOURCE RECOVERY DEVELOPMENT

Thomasine Hill Troisi[1] and John L. Thames

School of Renewable Natural Resources
University of Arizona
Tucson, Arizona

SUMMARY

Today Indian country is under siege. This threat derives from newly found natural resources on Indian lands: the minerals and gases beneath reservation lands, the water that moves through the earth, and the land itself. In their struggle to retain control over these resources, Indians face not only a formidable array of interested parties but the peculiarly modern problem of striking a balance between the development of these resources for their own benefit and the preservation of their ancient traditions (Kellogg, 1978).

INTRODUCTION

The wealth of the natural resources on Indian-controlled lands is immense and cannot be overestimated. America's Indian tribes control fully 16 percent of the nation's coal (including 30 percent of all coal west of the Mississippi) plus more than half of its uranium and about 3 percent of its still untapped oil and natural gas (*New York Times,* 1978).

The largest deposits of coal are in New Mexico and, primarily, Montana where an estimated 14 billion tons of low sulfur coal lie in the Fort Union formation; much of it is beneath the Crow and Northern Cheyenne reservations. When the boundaries of the reservations were established by treaty agreements in the mid-1800's, U.S. officials were unaware of the immense wealth that lay beneath that ground. Today, these reservations, especially the Crow, are targets of intense speculation by profit seeking power companies and a parade of natural resource consultant firms and high priced law firms. This chapter will discuss these and other pressures upon the Crow tribe which affect their attitude toward energy development.

HISTORICAL BACKGROUND

The Crow Indian reservation is in south central Montana about 60 miles southeast of Billings. It is the home of over 5,000 members of the Crow Indian tribe. About 1,500 tribal members live off the reservation; most of these live nearby and spend considerable time each year on the reservation.

Originally, the Crows were a part of the Hidatsa (Siouan) Tribe but in the mid-1700's two brothers, Red Feather and No Vitals, participated in a vision quest in which they were given the gifts of the Corn and the Sacred Tobacco Seed that led to the creation of the first Crow tribe (*Hardin Herald Guide to Big Horn County, 1978*). The first Crow people migrated from the midwest around 1776 to a huge 38 million acre area that occupied much of what is now Montana and Wyoming. They called this region Absaroka, the land of the Sparrow Hawk people (ibid., p. 15).

Early in their history, the Crows were an agricultural people but as their economy became dependent upon the buffalo, a culture evolved in which the highest recognition was obtained by military valor. The Crows were at war with other tribes competing for hunting territory during the period when the United States Army was attempting to settle Indians on reservations. The Crows joined with the U.S. Army in fighting other tribes. Their friendliness was rewarded in the Laramie Treaty of 1851 which established the boundaries of the reservation, giving the tribe more than 38 million acres, but in 1868 a second Laramie Treaty was signed that reduced the reservation to 9 million acres. Over the years, land cession to the United States, the Northern Pacific Railroad, the state of Montana, and sales to non-Indians have reduced the Crow holdings in trust to 1,567,348 acres.

CROW TRIBAL GOVERNMENT

The Crow Tribe did not accept the provisions of the Wheeler-Howard Act which created within each tribal government a small Executive Business Committee to conduct tribal affairs. Instead, in 1948, the Crow tribe, in what has been called a higher expression of inherent tribal sovereignty and governmental prerogative, organized its own Constitution and Bylaws; these are administered through the Crow Tribal Council (Schoppert, 1977, p. 5).

> The Crow Tribal Council: . . . the voice of the Crow Tribe; it is the medium, the body, the tribal organization through which the Crow Tribe speaks to the government and the general public; it represents the entire Crow tribe, it shall voice the opinions, wishes, sentiment, hopes and decisions in any and all tribal matters for the Crow people to the Congress and the Interior Department, by resolutions and through tribally elected delegates who shall, under instructions of the council, proceed to Washington or elsewhere to present in person such decisions and their own arguments and appeals in support thereof as the Council shall direct by majority vote (*Constitution and By-laws of the Crow Tribe,* Article VII).

Each tribal member, within certain age restrictions, is a member of the Crow Tribal Council and each member exercises one vote at each council meeting called (*Crow Constitution,* Article III). Regular Tribal Council meetings are held quarterly and special sessions are called if needed. Each meeting requires a quorum of 100 tribally enrolled members before the Council begins and attendance varies between 150 and over 1,000 members. This system of government is similiar to a "townhall" type of government or "pure democracy" which has often been criticized by outsiders as being cumbersome. Regardless of its size, it has been effective as a mechanism in involving the Crows as active participants in protecting their natural resources.

THE PRESSURES OF COAL DEVELOPMENT

The Crow nation is presently under an attack which could eventually lead to its destruction. The Crow tribe is forced into a position where it must decide to either develop its immense energy resource and attain significant wealth or not to develop it and thereby maintain the present economy in which many tribal people exist near the poverty level; the estimated unemployment rate for the tribe is 41.2 percent, more than five times the national average (Rocky Mountain Research Corporation, p. 49).

The source of the dilemma facing the Crow nation is 5 to 6 billion tons of mineable coal, located along the eastern edge of the reservation. Its development is imminent and could be the factor that guarantees a steady source of income for the tribe after years of economic instability (*Hardin Herald,* 1978). However, if this development is not done with care and wisdom, it could also destroy the Crow Nation's economy and culture. In January of 1975, the government released a Draft Programatic Environment Impact Statement on Crow Coal Development. It stated that:

"The adverse impacts would be the dilution of the Crow way of life through the increasing industrial activity. It could mean loss of cultural heritage and even the eventual extinction of the Crow language."

Presently, the Crow tribe is besieged by small time promoters, legitimate entrepreneurs, governmental pressures and industry. Each espouses a unique plan for instant wealth that will provide an answer to impoverished Crows. These individual entities have a remarkable effect upon the Crow coal attitude.

Promoters

A southwestern Indian approached the Crow tribe as a financial expert and consultant to a natural resources engineering firm that proposed a contract with the tribe to be the "middle man" or group to advise the tribe on mineral development with any outside interest. Initially some members of the tribe were impressed by his fast sales approach, friendliness, and capabilities in handling and breaking horses. Fortunately an administrative assistant to the then Tribal Chairman did a background investigation of the promoter and found the gentleman had no experience in coal negotiations, had operated an Indian jewelry shop and used car business, and had to resign from a governmental job because of incompetence. In addition, some tribal members became more cautious with the promoter when he was found to have taped conversations and important internal tribal meetings with recorders that were attached to his body. The tribe never signed a contract with the promoter and he has faded into the historical background of Crow coal development.

Entrepreneurs

Legitimate entrepreneurs come in all sizes and shapes with various amounts of money to offer. The present besieger has offered the tribe a one billion dollar proposal. Of course, he will receive his percentage if the proposal is consummated. The proposal is "to consummate a sale of all the coal reserves known to exist in the 200,000 acre coal belt which is not presently under litigation with various energy companies. It is estimated that this sale would yield at least one billion dollars within 2 years from commencement of the program" (Tornese, March 29, 1979, p. l). He has suggested that the "proceeds of the sale be deposited with a financial institution along with instructions to disburse interest only to the Crow tribe on an annual basis. This interest should be $80-90 million per year or approximately $45,000 per household. By preserving the $1 billion principal, these benefits would accrue to the Crow indefinitely and the financial independence of future generations would be guaranteed" (ibid., p. 1).

The members of the Crow tribe are divided over the billion dollar offer. Some Crows claim that the gentleman proposing the offer is not "a coal negotiator." They charge that if the tribe signs the agreement it would provide the entrepreneur with three million dollars off the top of the sale of coal royalties on the Westmoreland (the only company mining Crow coal in the ceded area) lease plus another eight percent commission. It is felt by some members that they don't need a "middle man" (*Hardin Herald,* December 14, 1978, p. 18).

The negotiator says he is a businessman and is "new to coal" but had been previously "in large scale money negotiations." Since he had not been hired by the Crow Tribal Council as its negotiator, he would only reveal his interested parties as a "domestic buyer rather than international" (ibid., p. 18).

Those members who support the billion dollar proposal see it as a means of providing needed income for tribal members and the tribe. In addition, there is a stressful matter which involves buying back land on the reservation owned by both tribal members and non-Indians who want to sell (*Hardin Herald,* December 7, 1978, p. 10) thereby maintaining tribal jurisdiction over tribal lands. Income from the coal would make this possible.

Federal Government and Organized Industry

The Crow Indians have been giving coal companies "headaches for years now, but the companies keep coming back" (*Hardin Herald,* August 24, 1978, p. 10A). This has been for a good reason since the Crows control 1/5 of all western coal and the companies want to develop it. However, there are many pitfalls that must be dealt with between the signing of a contract and the commencement of actual mining of the coal; these involve the various interested parties: the tribe, the federal government, Westmoreland, Shell, AMAX, Peabody Coal, and Gulf Oil.

Today, Westmoreland Resources is the only energy company mining Crow coal in the ceded areas here, the subsurface is owned by the tribe and the surface is owned by non-Indians or energy companies. Initially the

Westmoreland lease was to pay 17 1/2 cents per ton on 5 million tons a year, but since obtaining a revision of the Westmoreland lease in 1974, the tribe now receives royalties equal to 8 percent of the market value of the coal mined. However, the other four companies which have permits to lease within the reservation boundaries are still locked into a bottleneck of litigation, negotiation, and renegotiation with the tribe (ibid., p. 10A).

In 1974 the Tribal Council voted to void the energy companies' leases with the exception of Westmoreland whose lease had been renegotiated. The Council initiated their law suit, *Crow Tribe v. Secretary Kleppe,* that involved the four energy companies, the Secretary of Interior and the Commissioner of Indian Affairs, as defendants in the suit. The Tribe contends that the Bureau of Indian Affairs recommended "approval of the lease without the expertise required to make coal deals and before tribal members were familiar with how to negotiate with coal companies" (ibid., p. 10A). In addition, the allegations directed at the government state that (*KO:TTA':HILIK,* March 8, 1976, p. 8):

1. The permits and leases do not comply with the code of federal regulations.

2. The Secretary of Interior and Commissioner of Indian Affairs did not exercise their trust responsibility in issuing those leases and permits.

3. The leases and permits do not comply with the National Environmental Policy Act.

Even though the Crow Tribal Council filed the lawsuit in July of 1974, it was not until January of 1977 that Secretary Kleppe acknowledged the fact that officials of the Bureau of Indian Affairs, an Interior Department agency, had failed to protect tribal interest in negotiating the disputed strip mining agreements between 1968 and 1973 with the four energy companies. The Secretary's decision overturned a series of massive corporate leases of coal land on the Crow reservation which effectively blocked the planned strip mining of billions of tons of coal for years.

Secretary Kleppe had based his decision to revoke the leases on the failure of the Bureau of Indian Affairs "to observe or to obtain waivers of a long standing departmental regulation that forbid the leasing of more than 2,560 acres of any Indian land to a single party for mineral extraction unless it is 'in the best interest' of the tribe and a specific use is shown for the product, such as an adjacent electric power station" (*Billings Gazette,* January 14, 1977, p. 1). This decision stemmed from the fact that the energy companies exceeded the acreage limitation with Shell's lease covering 30,000 acres, AMAX's lease covering 14,000 acres, Peabody's lease covering 11,000 acres and Gulf's lease covering 70,000 acres. By reducing the four corporate leaseholds to a maximum of 2,560 acres each, 'a tract too small for economic strip mining,' the Secretary effectively moved the companies and the tribe back to square one and urged them 'to negotiate anew' (ibid., p. 10A).

Kleppe's administrative decision also ruled that "any new negotiation to broaden the companies' nominal coal rights not only would have to be conducted rigidly under regulations to protect rights, but also would require new environmental impact studies, a process that alone often requires more than a year to complete and can be challenged in the courts" (ibid., p. 10A).

On May 17, 1978 the tribe won a partial victory in the District Court ruling which found approval of the leases 'defective,' and granted 'partial summary judgement' to the tribe. Just what that ruling means is now being disputed between the tribe and the Bureau of Indian Affairs. The Crow Coal Authority, the Crow coal negotiating entity, contends that the ruling nullified the previous leases, while the Bureau of Indian Affairs attorneys believe that it only requires that they be renegotiated. Meanwhile the Crow Coal Authority continues new negotiations with AMAX, Shell, and other interested parties.

Minority Land Owners On Reservation

Besides possibly receiving billions of dollars from coal development, getting the federal government to acknowledge their failure in exercising their trust responsibility to the Crow tribe and then remedying it, why should the Crow tribe develop their coal? Primarily, they feel economic pressures to develop in order to establish economy control over their own tribal land base. Currently, many tribal members who are in a critical economic situation are requesting fee patents from the Bureau of Indian Affairs so that they can sell their land to non-Indians. Today, only 47 percent of the Crow reservation is owned by Indian people and held in trust while 43 percent is owned by non-Indians. The philosophy of the Crow tribe is to buy non-Indian lands on the reservation from non-Indian land owners who want to sell and eventually eliminate, as much as possible, all non-Indian land holders on the reservation and thereby maintaining these lands in perpetuity to the benefit of the Crow people. However, 'the current situation is the reverse' (Lozar, p. 1). Unfortunately, the tribe is not in a financial position to satisfy all the applications for sale, or to take advantage of the opportunities to buy available fee lands from non-Indian owners (ibid., p. 1). If the tribe is

> . . . in a position where money cannot be generated through loans, sale of coal, sale of other minerals, such as uranium, or oil and gas in sufficient quantity to satisfy the request for sale, then this inability

will cause the Crow reservation land base to shrink. The less land that the Crow tribe has and the less land that the Crow individual has, will eventually cause them to become the minority land owners on their own reservation (ibid., p. 2).

In order for the tribe to achieve economic self-sufficiency and to be the major land owner on the reservation, it will need to wisely and carefully develop its coal without harming its culture and destroying the environment. However, to attain this independence the Crows must plan for a large overall purchase of all lands that are available for sale by Indians and, especially, non-Indians who are desirous of selling their land. The Crows realize that their reservation must become one which will support the working family and, if the tribe can be in a position with sufficient funds to buy non-Indian land operators out, it is conceivable that the Crows could become "the largest and most effective cattle ranchers in the West" (Lozar, p. 2).

CONCLUSION

It is difficult to speculate upon how the Crow Tribal Council will eventually resolve the coal issue. From their legal battles with the federal government and energy companies and their internal power struggles to determine which political factions will be the controlling decision makers, it is apparent that the Crows are a proud people with a strong individualistic nature which has served them well in external battles with off-reservation and non-Indian entities, but which has also torn at the fiber and soil of the tribe in internal struggles. Aside from these struggles, there is the dark reality of possible financial loss to outside energy companies if the initial coal negotiation and final agreements are not economically satisfactory or jurisdictional loss if their impoverished tribal members, who live from day to day, need to sell their land to wealthier non-Indians in order to survive. Will the Crow tribe survive these threats upon its homeland? Survival as a people is not foreign to the Crow. They are shrewd and see their energy resources as a means of strengthening their sovereignty and thereby preserving their culture. The interesting part is yet to come. After the experience of almost being sold out by the federal government and bought out cheaply by the energy companies, the people of the Crow tribe can now sit face to face with representatives of the billion dollar energy firms and let them know what they want. It is to be hoped that this will be a new era for the Crow people.

NOTES

1. Thomasine Hill Troisi is a member of the Crow Tribe and has been active in tribal affairs.

REFERENCES

"Billion Dollar Coal Deal to Crow Council," *Hardin Herald,* December 7, 1978, p. 1.

"Coal: The Last Chance for the Crow," *The New York Times,* January 8, 1978, Section 3 p. 1, 11.

Constitution and By-Laws of the Crow Tribe.

"Crows Give Coal Companies a Headache," *Hardin Herald,* August 24, 1978, p. 10A.

"Crows Hurl Charges Like Artillery Barrage," *Hardin Herald,* December 14, 1978, p. 18.

"Crows Refile Lawsuit After Bureaucratic Stalling Tactic," *KO:TTA':HILIK,* March 3, 1976, p. 18.

Draft Programmatic Environmental Impact Statement On Crow Coal Development, U.S. Department of Interior, Bureau of Indian Affairs. Billings, Montana, January, 1975.

"Early Crow History Affected By Gold, Other Indian Tribes," *Hardin Herald Guide to Big Horn County,* 1978, p. 15.

Kellogg, M. "Indian Rights: Fighting Back With White Man's Weapons," *Saturday Review,* (November 25, 1978), p. 24-27.

"Kleppe Tosses Out Crow Coal Leases," *Billings Gazette,* January 14, 1977, p. 1, 10A.

Lozar, S. "Letter to Mr. Forest Horn, Crow Tribal Chairman," Unpublished letter, U.S. Department of Interior, Bureau of Indian Affairs, Crow Agency, Montana, February 27, 1979.

Rocky Mountain Research Corporation. *Crow Tribe Resource Development and Land Use Plan,* Provo, Utah, October, 1977.

Schoppert, T. K. "Designation Request by the Northern Cheyenne Tribe." Unpublished letter, Billings, Montana: Law Offices of Lynaugh and Fitzgerald, June 29, 1977.

Tornese, T. R. "Letter to Mr. Forest Horn, Crow Tribal Chairman," Unpublished letter, Washington, D.C., March 29, 1979.

4

ENVIRONMENTAL IMPACTS OF ENERGY RESOURCE RECOVERY IN ARID LANDS

Cyrus M. McKell

Institute for Land Rehabilitation
Utah State University
Logan, Utah

SUMMARY

The arid and semiarid West has abundant energy resources. Yet in recovering them drastic environmental disturbances occur. The important point in assessing environmental impacts of recovery operations is not that physical disruption to specific aspects of topography, soil, underground and surface hydrologic features occurs or that plants and animals are displaced or destroyed but that ecological systems are disrupted or terminated. Philosophical conflicts between energy development versus environmental protection have been acute but will continue to generate controversy as we deal with such problems as impact mitigation, assignment of costs, and commitment of scarce western resources.

INTRODUCTION

The accident at the Three Mile Island nuclear power generating plant on March 28, 1979 illustrates the growing dilemma of decision makers and citizens in general regarding ways to fill the power gap in the nation's expanding energy needs while at the same time maintaining environmental quality and safety.

Increased recovery and development of abundant energy resources in the arid and semiarid western states could help to fill the gaps in national

41

energy production but not without incurring substantial costs. Coal was seen to be especially important as a replacement for fuel oil and natural gas in the President's National Energy Plan presented in 1977 and enacted in less stringent form by Congress in 1978 (Council of Environmental Quality 1978). Many of the costs for increased energy development are economic and could be included in the price of the energy product but the environmental and social costs are more difficult to assess and collect—not to mention the ability to mitigate the impacts that would be created.

The urgency of meeting energy needs was declared in President Carter's message to the country on July 15, 1979 in which he proposed a National Energy Development Corporation to stimulate and underwrite a program of synthetic fuels development and a super agency, the Energy Mobilization Board to cut through government red tape and regulations to expedite high priority projects.

Environmental leaders are alarmed that carefully laid procedures for identification, resolution or avoidance of environmental impacts will be ignored in favor of full scale development. Western governors expressed concern in their meeting with President Carter on October 10, 1979 in Albuquerque, New Mexico, that state's rights and environmental quality would be protected in any procedure for "fast track" decision making (Pierce 1979). On the other hand, energy production interests see in the Energy Mobilization Board a mechanism to avoid unreasonable delays in developing much-needed resources. Where the emphasis will be placed regarding these two conflicting positions will partly depend on the mood of the nation which appears to be shifting in favor of increasing the supply of energy by accepting certain levels of environmental impacts, especially if they can be adequately mitigated.

About 24 percent of the eight Rocky Mountain states (Arizona, Colorado, Montana, New Mexico, North Dakota, South Dakota, Utah and Wyoming) is underlaid by coal (Averitt 1974). Other energy resources such as oil shale, uranium, petroleum and natural gas, tar sands, geothermal and hydro are also abundant as discussed by W. H. Dresher in his review of energy resources located in arid lands (Chapter 1).

Of the various environmental impacts that would occur in developing western energy resources, those inherent in coal mining, especially surface mining, (Ralston et al. 1977) would be the most significant and are emphasized in this chapter.

Recovery and development of coal and other energy resources have not been without problems and impacts in the past and may be expected to be more difficult in the future. The National Energy Plan directed that a review be undertaken of the possible consequences of a rapid increase in coal production. The report of a study committee chaired by P. Rall of the National Institute of Environmental Health Sciences, reported that a safe doubling[1] of coal use by 1985, which would involve producing 1.2 billion tons of coal, would require rigorous adherence to five principles:

    1. Compliance with all air, water and solid waste regulations.

    2. Universal adoption and successful operation of best available technology in new facilities.

    3. Compliance with reclamation standards.

    4. Compliance with mine health and safety standards.

    5. Judicious siting of coal-fired facilities.

From a national point of view these principles are obviously important and justifiable. But in the arid western states, considerably more specificity is necessary to avoid environmental damage. Three issues are of paramount concern and must be addressed in any policy or energy resource development land in the arid west:

    1. Water is scarce.

    2. Scenic values are high.

    3. Arid lands are fragile.

Water Resources

Water resources in the Intermountain States are already committed to agricultural, urban, and industrial uses and any new uses for energy resource development will have to compete with the present users for this limited resource. Although energy resource recovery will require negligible amounts of water, substantial quantities would be needed for energy conversion and transportation (Davis and Wood 1974) plus urban development to service the expanded energy industry. Whether such changes in water use would be congruent with the economic and social goals of western states is a serious concern that must be addressed.

High Scenic Values

High scenic values in the western states represent a national asset and regional heritage. Clear skies, undeveloped vistas and wide-open spaces should not be marred by needless or shoddy development. Twelve national parks plus many more national monuments make the seven-state intermountain region one of the great scenic areas of the nation. High environmental quality is a necessary backdrop for the scenic grandeur not only of the established parks and monuments but also for the towering mountains, deep canyons, desert vistas and verdant valleys.

Fragile Nature of Arid Lands

The fragile nature of arid lands is based on unique land characteristics: shallow and infertile soils (Dregne 1963), a sometimes sparse plant cover (Kuchler 1964) and extreme fluctuations and intensity of weather factors (McGehee 1964, Meigs 1960) such as short duration storms, hot summers and cold winters, wind storms, and high summer evaporation. Each of these extremes plus many others create problems of soil instability and erosion, limited plant growth, low productivity and high hazards for plant and animal reproduction and survival. The resulting plant communities (Kuchler 1964, Cronquist et al. 1972) are adapted to the harsh soil and climatic conditions.

Drastic disturbances to the soil and underlying strata are difficult to remedy. Rehabilitation is an answer to surface disturbance but regeneration of plant or animal populations takes time in arid regions. Under natural conditions many years may pass until a favorable sequence of events allows the establishment of a stable plant community. Rehabilitation measures face a formidable challenge in overcoming unfavorable weather conditions but also in dealing with immature and heterogeneous soil or spoil as a plant growth medium.

ENVIRONMENTAL IMPACTS

In discussing environmental impacts the point should be made that merely identifying an impact in an EIS (U.S. Dept. Interior 1974, 1976, 1979) does not necessarily indicate its importance. Further, there are large differences in the costs and methods available to mitigate impacts. These considerations should be examined carefully in assessing the importance or severity of environmental impacts resulting from energy resource development in arid lands.

Impacts to Physical Environment

Basic to the biological environment and its functions are the physical or abiotic components of an ecosystem. These components include air, water and soil and each is subject to some degree of impact from mining, depending on the methods used, and the magnitude of the operation.

A concept that is often overlooked when considering impacts is that the recovery (mining) operations are only a temporary use of the land (Leathers 1977). This is emphasized in the general philosophy of the Surface Mining Control and Land Reclamation Act of 1977 which requires that land be returned—"to a condition capable of supporting uses which it was capable of supporting prior to any mining or, higher or better uses—."

Air. Impacts to air quality are caused mostly from dust created in such mining operations as spoiling, hauling and blasting. Dust occurrence is local when associated with specific operations and can be mitigated to some degree by water sprinkling, oiling, or graveling of roads. A more general dust problem is created where extensive spoil piles are left unreclaimed in areas of high wind. In such cases, visibility would be continually impacted. Various environmental impact statements for coal development projects attempt to describe this problem (U.S. Dept. Interior 1979).

Air quality problems are increased when power plants are sited near surface mining operations. Even with air pollution abatement equipment as mandated by the Clean Air Act, power plants can have far reaching impacts and may have to be sited at some distance from the coal source if Class I air quality areas are nearby. Shifting the 3,000 megawatt capacity Intermountain Power Project plant from the proposed site 8 miles east of Capitol Reef National Park to a new site in the Great Basin at Lynndyl, Utah is an example of such an accommodation.

Noise may be listed as an impact of the air resource. For the most part it is caused by heavy equipment operations that have only local influence. However, regular blasting to fracture overburden or coal seams is a more serious impact and it may affect wildlife migration or reproduction in adjacent areas.

Water. In the arid west, impacts to water quality or quantity cannot be taken lightly. Provisions of the Clean Water Act of 1977 and earlier water pollution laws set water quality standards for use and discharge. Operations which extract energy resources can have significant influences on both the amount and quality of water. Surface mining can directly impact water quantity by diverting surface flow, intercepting underground aquifers and dewatering wells and springs. Utilization of water for onsite processes and land rehabilitation as well as for increased community use by the mining work force may well require the purchase and withdrawal of water from uses that existed prior to mining, thus extending the impact to the surrounding region. A large number of regulatory provisions are contained in the Surface Mine Control and Reclamation Act of 1977 and various state mining laws regarding the general disruption of the hydrologic system.

Water quality reduction is a natural consequence of soil and geologic strata disturbance. An increase in dissolved and suspended solids occurs when overland flow and erosion flush soil particles into water systems. Solutes of various kinds may diffuse into water from toxic concentrations in the overburden. Salts from saline topsoil may be leached into the runoff should such topsoil be replaced in response to rehabilitation requirement. Off-site impacts from water of reduced quality can be minimized if off-site water flow is prevented. When rainfall is limited, off-site flow may not be a problem but where some flow occurred prior to mining, the problem of containment on site versus depriving water from down stream users would cause serious social and environmental impacts.

Land and Topography. Impacts to the surface configuration of the land are the most obvious consequences of energy resource removal. Since most energy recovery activities would be disruptive to ecosystems of high environmental sensibility, they would be precluded in areas designated as wilderness, wild and scenic rivers or Class I air quality areas by the respective laws governing such designations. Mining and exploration cause topographic changes by creating mining pits, spoil piles and exploratory drilling sites. Prior to the passage of state mining laws in the 1970's, spoil piles, final pits, and highwalls remained as ugly scars in the landscape after mining. Under PL 94/87 and state laws, topographic relief following mining must now be rendered similar to the surrounding area unless leaving pits and high walls would serve a better use such as enhancement of wildlife habitat by employing pits to create a permanent water source and aquatic diversity.

Subsidence, slumping, and erosion are impacts that have occurred in the past but under present regulations must be prevented by appropriate compaction, regrading and surface protection.

Resource recovery that requires the removal of soil and overburden causes the most significant impact to an ecosystem and its productivity. Soil structure, horizon stratification, soil moisture and drainage relationships, nutrient cycling and microbiological populations, to name a few of the important functions, are drastically altered if not destroyed in the mining process. Current regulations call for the replacement of topsoil and overburden as the first step in reclamation. However, this has been described by one reclamation soil scientist (Grogan 1979) as like pushing a barn over with a bulldozer and then pushing the pieces back onto the site and expecting it to function like a barn again! The essential ingredient, time, must also be allowed for the soil functions (analogous to the barn pieces) to become reestablished. We must be reminded that in H. Jenny's (1941) formula for the formation of soil, time is a major component.

Impacts to Biological Environment

Plants. Individual plants making up the vegetation of an area receive a direct impact as a result of mining and energy resource exploration. Disturbance ranges from total destruction of plants on a surface mine site to stem damage from vehicle traffic in exploration and drilling sites. Deep topsoil removal may destroy the compliment of seeds from a plant community but seeds from a shallow topsoil layer may be very useful to initiate natural revegetation of disturbed areas (Beauchamp et al. 1975). Plants remaining in areas adjacent to disturbed sites may receive a secondary impact from dust, surface erosion and runoff, and increased wildlife use by displaced animals.

Revegetation operations undertaken to replace the plant cover of the soil face great obstacles in deciding the appropriate species mix and how to obtain seedling establishment under arid conditions. Multiple use objectives are not easy to achieve in providing a stable plant cover, a suitable wildlife habitat and a sufficient amount of biomass to satisfy the premining land uses.

Plant species that are endangered because of their scarcity and restriction to critical habitats are now protected from impact by the Endangered Species Act. The dearth of habitat and distribution information is a deterrent to effective decision making with regard to use of specific sites where energy resources occur.

Animals. Even though many animal species have the ability to migrate when they are threatened by impacts of mining operations as parts of their habitat are destroyed. Loss of food and cover plants, soil, landform and water plus the disturbing activity of mining cause animal species to move to adjacent areas where they would compete for habitat factors with residents there. If they can't migrate, destruction is their fate. Reestablishment of animal populations is not just a matter of migration back into a reclaimed area. The ecological balance cannot be established in the first stages of revegetation because plants may not be sufficiently abundant, may not be in ecologically balanced proportions, or cannot withstand a substantial loss of tissues or seeds to animals. Time, as with other functions, is required to achieve a degree of stability in animal numbers and species in relation to the carrying capacity of the reclaimed habitat.

Areas of critical habitat for endangered animal species can only be used for energy resource development if that activity is consistent with survival of the endangered species.

Soil Micro Organisms. Disturbance or destruction of soil creates a drastic impact to soil fauna and flora. Replacement of topsoil, as required by regulation is at best a means of reinnoculating the heterogeneous surface layer that results. Stock piling topsoil when it cannot be used immediately has been shown to create some problems (Miller and Cameron 1976) by shifting the center area of the topsoil pile to anaerobic condition, resulting in the loss of a major portion of the soil biological activity. In arid regions, the degree of biological activity in the top layer of soil is probably at a minimum and only exists in any abundance in the "islands of fertility" (Garcia-Moya and McKell 1972, West and Skujins 1979) near the base of shrubs.

Only recently has it been shown (Aldon and Springfield, 1975) that mycorrhizael innoculation is important for plant establishment and survival on sterile soils or spoils.

Basis for Identification of Impacts

Concern for the harmful effects of activities that degraded environmental quality led to passage of the National Environmental Policy Act of 1969. Although the NEPA act made the words "environmental impact" well known to people in general, two of the most important results were to require each federal agency to consider environmental protection as a part of its already existing mission function and to require an agency to analyze its significant actions for any environmental impacts that would occur if the proposed action were implemented. Preparation of an environmental impact statement (EIS) was the agency's output in discharging its environmental responsibility. Adequacy of the EIS could be challenged in court, and many were, but the judgement of the agency regarding its response to the identified impacts could not be challenged according to NEPA provisions. Environmental impact statements increased in encyclopedic volume in order to adequately cover all aspects of the various issues. The result was not better clarity but confusion and delay. In June 1978, new NEPA regulations were issued which require the EIS process to "scope" the various impacts and select only the important ones for concentrated study.

Partly as a result of the inadequacy of NEPA to force a specific action and partly because there was a need for impacts of critical concern to be identified and limited, numerous environmental laws were passed or amended which set pollution limits, protected critical environmental resources and specified practices to be used in managing critical resources.

Major environmental laws and their requirements now provide the basis for defining the limits of some environmental impacts (Table 4.1).

Table 4.1. Categories of environmental impact, federal laws to control impacts and some control functions of the law

Category	Federal Law and Some Performance Requirements
Air Quality	Clean Air Act Amendments of 1977.

Set standards for:
- Photochemical oxidants, less than 0.08 ppm.
- Total suspended particulates, under study.
- Sulfur oxides, insufficient data to set a limit at this time.
- Lead, EPA recommended not more than 1.5 mg/m³.
- Carbon monoxide, 10 mg/m³
- Nitrogen dioxide, short-term standards set in California at 470 mg/m³. (Other areas to be set as data become adequate).
- Pollution standard index, daily maximum pollution levels of 100 (air quality control regions to adopt this limit by 1978).
- Established criteria for Class I, II and III air quality areas.

| Water Quality | Federal Water Pollution Control Act of 1972. Clean Water Act of 1977. |

- Revised certain requirements of the 1972 act, established 1983 as a deadline for water quality compliance.
- Identified a list of toxic pollutants and set limits and deadlines for meeting limits.
- Discharge of "non conventional" pollutant to be processed by best available technology.
- Restrict dissolved solids in irrigation return flow.

| Solid Waste | Resource Conservation and Recovery Act of 1976. |

- Control of hazardous wastes.
- Land disposal of solid wastes controlled.

| Natural Resources | National Forest Management Act of 1976. |

- National forests must be managed to meet a broad range of publicly beneficial goals.

Resources Planning Act of 1974.

- Agencies to develop land and resources management plans to integrate and control uses.

| Wilderness | Wilderness Act of 1964 and 1978. |

- Designated 16.5 million acres of land with wilderness character to be managed under limited use.

| Endangered Species | Endangered Species Act of 1973. |

- Protection of habitat critical to the survival of endangered species.

| Wild and Scenic Rivers | Wild and Scenic Rivers Act. PL 90-541 |

- Restricts development of rivers so designated.

| Surface Mining | Surface Mine Control and Reclamation Act of 1977 |

- Restricts mining of unsuitable areas.
- Sets environmental standards for reclamation.
- Provides funds for reclaiming abandoned lands.

Ecological Interrelationships Must be Considered

Lacking in many Environmental Impact Statements is a careful discussion of the interrelationships among ecological factors and the way in which a change in one environmental factor will result in a chain reaction involving other factors. The significance of a given impact cannot be correctly assessed unless its relationship to other functions is known. For example, to mitigate the low fertility of material substituted for topsoil in a rehabilitation program, the soil material is heavily fertilized. The result can be a drastic but temporary increase in annual weedy plant species which in turn utilize the added fertilizer and cause excessive competition with the perennial plant species for soil moisture and reduce the survival of the perennial species. The net result may be a reduced and weak stand of perennial native plants.

In addition to the environmental impacts of energy resource development there are cultural components which are beyond the scope of this chapter. Nevertheless, impacts to the cultural environment are important and include such concerns as archeological site and historical landmarks that may be destroyed, land use patterns that would be altered or given up, aesthetic values that would be compromised, recreation opportunities foregone, social environment and lifestyle subjected to drastic change, and economic aspects of communities changed both by increased opportunities for income and by greater costs for community development.

Each environmental impact statement must include a discussion of the opportunities to mitigate the expected impacts. Some possibilities for mitigation can be implemented easily and at a low cost such as changing a location or a sequence of operations. Other measures can be implemented only by purchase of expensive equipment and major treatments over a long period of time. Whatever the mitigation strategy, if it will work and is cost effective, it should be used. In the same context, there should be a willingness to accept mitigation measures as a solution to impact abatement and not refuse development on the philosophical grounds that "no development" is a better long term alternative.

However, there are certain impacts that cannot be mitigated completely. These impacts are listed in a separate section of the EIS and include the use of resources in the development process, temporary loss of vegetal cover and wildlife habitat, changed land use patterns, modified aesthetic values and new socio-economic conditions. Whether these impacts are serious enough to preclude project development is a question that has been solved in the political arena by examining the trade-off values. At the present time, achieving energy sufficiency is a powerful incentive to consider in the decision-making process.

ASSESSING THE IMPORTANCE OF ENVIRONMENTAL IMPACTS

One of the major problems in making an environmental study is assessing the importance of environmental impacts listed in the reports. Many impacts can be quantified in some way or another so that comparisons can be made against other impacts or against the consequences. However, some impacts cannot be quantified nor can the consequences of ignoring them be fully predicted because the time consequence is too long, data are lacking, or the relationships are not clear.

With regard to the impacts of energy resource recovery in arid lands many serious questions have been raised in EIS's for various coal fields (U.S. Department of Interior 1974, 1976, 1979).

After the impacts have been identified and described a set of questions may be asked in assessing their real importance.

    1. What is the duration of the impact and how much will it affect other environmental factors or areas?

    2. What is the present ability to mitigate the impact?

    3. What is the relative cost of mitigating the impact either in terms of the energy resource extracted or in terms of beneficial use of the area after mining.

An overall example of the application of these three questions is in the cost of rehabilitation after surface mining for coal. Leathers (1977) gathered data from most of the coal surface mined areas of the U.S. He assumed that annual income foregone and/or off-site pollution abatement costs amount to $25 per acre of unreclaimed land, thus, an estimate of the cost of present total direct damages would be about $13 million annually. Fortune (1975) estimates public costs of a comprehensive, nation-wide reclamation program for these "orphaned" lands at $385 million (in 1976 dollars). In annual terms the direct costs amount to approximately $60 per acre or more than double the "assumed" benefits. In the arid West, assumed benefits from nonagricultural land use would be far less than the nationwide average of $60 as given by Fortune.

SUMMARY

Many different kinds of impacts are created in the recovery of energy resources from arid lands. Air quality may be degraded by the dust from mining activities or vehicle traffic. Water quality can be reduced by runoff and erosion from disturbed or denuded areas. Some water may be used in various energy recovery operations but much larger quantities are required in energy conversion. Highly visible impacts to the soil and geologic strata may occur from mining, drilling and exploration.

Impacts to biological systems can be far-reaching and time is required to restore plant and animal populations to stable conditions. Impacts to microbiological systems are not well understood but ways are being developed to restore their functions and obtain the benefits they provide to ecosystems.

Some impacts are obviously more serious than others even though it is not always possible to quantify them or obtain general consensus as to their importance. There is a lack of agreement as to the values that are involved if the impacts are not mitigated but it is biologically impossible and economically unreasonable to mitigate all of the impacts.

We all want cheap energy to sustain our present life style. A number of policy issues are immediately evident as the nation seeks to solve the problems of energy resource development.

Which has a higher priority, energy or environment is a question that arises often but must be considered on a site-specific basis. Western governors assert that both state's responsibilities and environmental quality must not be abridged by any national goals or programs for energy sufficiency.

NOTES

1. This assumption of safe doubling would involve a production level of 1.2 billion tons of coal.

REFERENCES

Aldon, E. F. and H. W. Springfield. 1975. "Problems and Techniques in Revegetating Coal Mine Spoils in New Mexico." p. 122-132. *in* Wali, M. K. (ed) *Proc. of Symposium on Practices and Problems of Land Reclamation in Western North America.* Univ. of North Dakota Press, Fargo, North Dakota.

Averitt, P. 1975. "Coal Resources of the United States, January 1974." *Geological Survey Bulletin* 1412.

Beauchamp, H., R. Lang and M. May. 1975. "Topsoil as a Seed Source for Reseeding Strip Mine Spoils," *Wyo. Agric. Esp. Stn. Res. J.* 90.

Council on Environmental Quality. 1978. The Ninth Annual Report of the Council on Environmental Quality. U.S. Govt. Printing Office. Washington, D.C. 599 pp.

Cronquist, A., A. H. Cronquist, N. H. Holmgren and J. L. Reveal. 1972. Intermountain Flora, Vascular Plants of the Intermountain West, USA. Vol. 1. Hafner Pub. Co. New York. 270 pp.

Davis, G. H. and L. H. Wood. 1974. Water Demands for Expanding Energy Development. *Geological Survey Circular* 703. Reston, Virginia. 14 pp.

Dregne, H. E. 1963. "Soils of the Arid West." p. 215-238 *in* Hodge, C. and P. Duisberg (eds) *Aridity and Man.* Pub. 74 A.A.A.S., Washington, D.C.

Fortune, M. 1975. "Environmental Consequences of Extracting Coal." *in* Commoner, B. Bakesbaum and Corr (eds) *Energy and Human Welfare—A Critical Analysis* Vol 1: The Social Costs of Power Production. Macmillan Pub. Co., New York.

Garcia-Moya, E. and C. M. McKell. 1970. "Contribution of Shrubs to the Nitrogen Economy of a Desert Wash Plant Community." *Ecology* 51:81-88.

Grogan, S. 1979. Personal communication with reclamation supervisor, Utah International Navajo Mine.

Jenny, H. 1941. *Factors of Soil Formation.* McGraw-Hill Book Co., New York.

Kuchler, A. W. 1964. "Potential Natural Vegetation of the Conterminus United States." *Am. Geog. Soc. Soc.* Spec. Publ. 36. 154 pp.

Leathers, K. L. 1977. "The Economics of Land Reclamation in the Surface Mining of Coal: A Case Study of the Western Region of the United States." Paper No. 39. Natural Resource Economics Div. SEA-Economics Rec. USDA. 147 pp.

McGehee, R. M. 1964. "Weather: Complex Causes of Aridity." p. 117-143 *in* Hodge, C. and P. Duisberg (eds) *Aridity and Man.* Pub. 74. A.A.A.S. Washington, D.C.

Meigs, P. 1960. "Distribution of Arid Homoclimates." United Nations map no. 303 Western Hemisphere. United Nations, New York.

Miller, R. M. and R. E. Cameron. 1976. "Some Effects on Soil Microbiota of Topsoil Storage During Surface Mining" *Fourth Sumposium on Surface Mining and Reclamation.* National Coal Association and Bituminous Coal Research Inc. p 131-139. Louisville, Kentucky.

Pierce, N. R. 1979. "Savvy Governors Pool Energy." *The Salt Lake Tribune* Oct. 21, p 20.

Ralston, S., D. Hilbert, D. Swift, B. Carlson, and L. Mengies. 1977. The Ecological Effects of Coal Strip Mining—A Bibliography with Abstracts. US Dept. Interior, Fish and Wildlife Service, Western Energy and Land Use Team, Ft. Collins, Colorado.

U.S. Dept. of Interior. 1979 Final Environmental Impact Statement. Development of Coal Resources in Southern Utah. Bureau of Land Management.

U.S. Dept. of Interior. 1976. Final Environmental Impact Statement. Northwest Colorado Coal. Bureau of Land Management.

U.S. Dept. of Interior. 1974. Final Environmental Impact Statement. Eastern Powder River Coal Basin of Wyoming. Bureau of Land Management. 6 volumes.

West, N. E. and J. Skujins (eds) 1979. "Nitrogen in Desert Ecosystems. US/IBP Synthesis Series 9.

5

LEGISLATIVE REQUIREMENTS FOR
THE DEVELOPMENT OF ENERGY RESOURCES[1]

James D. Mertes

Department of Landscape Architecture
and Park Administration
Texas Tech University
Lubbock, Texas

and

Frank F. Skillern

Texas Tech University School of Law
Lubbock, Texas

SUMMARY

A discussion of the legislative requirements for the development of energy resources is inherently controversial. It should be undertaken in light of the objectives being sought through legislation. A principal goal is to minimize the negative external impacts that can result from mineral development. Objectives sought include increasing production of resources to meet a rapidly increasing demand for energy, providing consumers a product at the lowest possible price, preserving competition within an industry or for a resource by minimizing dominance by one company or group of companies, and protecting and preserving the physical environment.[2] The public is directly interested in most of these objectives, particularly when the mineral development is occurring on public lands.

Frequently statutory requirements and regulations evidence this public interest. In particular, the public interest is seen in requirements for notice, public hearings, and opportunities for various positions to be stated in public meetings held by the agency decisionmakers.[3] But the legislative concern for the public interest does not stop at merely assuring public participation in the agency decisionmaking process. It is also protected through direct access to the courts.

INTRODUCTION

With increasing frequency since 1970, legislation relating to the environment and development of energy resources has expressly authorized citizen suits to enforce the standards of the legislation.[4] A citizen suit provision allows a private party to enforce the statute which the party could not otherwise do.[5] Often this right of enforcement is independent of and in addition to the administrative requirements for public hearings and public proceedings in which the private parties may also participate. It is a right to seek judicial review of the procedures followed and the substance of the agency's decision after it has been made.

It is not our purpose to resolve the controversy over the merits of either the statutory requirements or the public involvement in agency decision-making. The purpose is not to evaluate the merits of these issues, but rather to present an overview of the statutory requirements, highlighting their content and their operation with respect to the development of various energy resources.

LEGISLATIVE REQUIREMENTS

Legislative requirements affecting the development of energy resources exist at the state and federal level and operate in a variety of ways. For example, they may arise through seeking permission in the form of a license from a governmental agency to undertake an activity. Or they may exist in the form of performance standards imposed under regulatory statutes such as the federal Clean Air Act[6] or federal Clean Water Act.[7] Another regulatory statute regulating health and safety conditions for workers is the federal Occupational Safety and Health Act.[8] The statutory mandates may also take the form of land use planning requirements. Examples of these include facility siting statutes or the National Environmental Policy Act of 1969 (NEPA)[9] which establishes certain procedures that must be fulfilled before an activity can be undertaken. Additional limitations on land use may be imposed on the development of resources in critical areas. This concern is often reflected through statutes such as the Wilderness Act,[10] the Coastal Zone Management Act,[11] and the National

Forest Management Act[12] which impose limitations or conditions on the types of activities that can be undertaken in critical areas or require permission of the agency managing the area before undertaking the activity there.

How legislative requirements affect development of mineral resources may be clarified by examining those requirements at different stages of developing the resource. The types of limitations and how they operate can be examined in the context of exploration, extraction and production, and transmission stages of development. Some of the statutory requirements may apply at each stage; others may apply only at a particular stage. This examination of the requirements at each stage will be followed by a discussion of the special issues concerning the use of public lands which will be treated separately. The administration of the federal statutes and various state approvals to these issues will be presented last.

EXPLORATION STAGE

At the exploration stage the principal statutes will be licensing and planning statutes. At the federal level in order to conduct exploration activities to determine if resources are available in commmercial quantity, it is necessary to get the permission of the agency managing the public lands.[13] Much of the coal lands in the west, for example, is under the management of the Bureau of Land Management in the Department of Interior or the Forest Service in the Department of Agriculture. Rights to drill for the oil and gas in Outer Continental Shelf are leased by the Department of Interior. Mineral leasing is done under the Mineral Leasing Act of 1920,[14] the Outer Continental Shelf Lands Act,[15] and the Federal Coal Leasing Act Amendments of 1976.[16] These statutes set forth the conditions under which a person may undertake prospecting or mining on federal property. The permission granted may be conditioned on whatever terms the agency deems reasonable and necessary. The applicant may have to establish financial solvency, to show technological capability, and to provide detailed plans for development.[17]

The exploration decision requires consideration of information concerning the ramifications of development in the future as well as of the immediate availability of the mineral. One purpose in the exploration stage is to predict whether the proposed development in the long run will cause environmental or other harm or will be in violation of applicable statutes. If so, the preliminary, exploratory work should be prohibited or allowed only under adequate safeguards to minimize this potential harm. In particular, attention must be given to the site of the activity to determine if it is in a critical area that will be protected under planning statutes such as the Coastal Zone Management Act of 1971 (CZMA)[18] or in an area such as a wilderness study area that will otherwise require protection under the Wilderness Act.[19] The Endangered Species Act[20] may also apply by requiring preservation of critical habitats of endangered or threatened species.[21] NEPA may necessitate development of mitigation strategies or use of an enviromentally less damaging alternative in the exploration stage. In addition, it must be determined if the activity will pose a threat to surface water in violation of standards under the federal Clean Water Act (CWA)[22] or create conditions that may violate the federal Clean Air Act (CAA).[23] Under the CAA, for example, if the activity is being undertaken in a pristine, clean air area, prevention of significant deterioration standards must be met.[24] If the proposed development is near a mandatory Class I area (which includes national parks and wilderness areas over 6,000 acres by August, 1977), the visibility requirements of the CAA must be considered.[25]

The visibility standards go beyond regulation of dust and particulate matter to the preservation of air quality related values in the area which may include scenic and aesthetic values as determined by the federal land manager. The federal land manager has an affirmative duty under the CAA to assure that these air quality related values are maintained.[26] That duty may include denying a permit for development in the area or vetoing one for development outside the area that would impair those values within the area. The extent of this authority and its use are currently being developed.

If the development is being proposed in a nonattainment (dirty air) area, then the licensee may have to find other sources of pollution to trade off and diminish their pollution before the development or activity can proceed.[27]

The exploration stage must also consider whether an environmental impact statement (EIS) must be prepared. NEPA requires that an EIS must be prepared for any major federal activity that significantly affects quality of the human environment. Exploration work may be delayed pending completion by the agency of a site specific EIS. Whether the EIS is required depends on the area, the nature of the action being undertaken, and its effect on the environment. NEPA has been treated expansively by the courts and clearly applies to federal agency action such as licensing mining activities, licensing construction of nuclear power plants, or funding certain activities which may be involved in exploration activity.

EXTRACTION AND PRODUCTION

At this stage of developing a resource several federal statutes may be immediately applicable. NEPA may apply to require that the least environmentally offensive means of extraction be used. In the course of requiring consideration of alternatives, NEPA may necessitate looking to alternatives that are environmentally more suitable and may require using all reasonable mitigation efforts. The Council on Environmental Quality has promulgated recent regulations that clarify this responsibility.[28] Under these regulations, if the environmentally preferable alternative is not the selected one, the agency must explain why the former was not chosen.[29]

This stage also brings into play performance standards under federal statutes. The federal CAA may apply to regulate the amount of dust and particulate matter that can be generated by the activity, or depending on the area, it may require compliance with prevention of significant deterioration (PSD) and visibility standards.

Regulatory standards under the federal CWA would also be directly applicable at this stage of development. The residue of drilling, brine water, leaching, acid mine water, and other pollutants could not be discharged into surface or ground water supplies without getting the necessary permit from the state National Pollutant Discharge Elimination System (NPDES) agency or the Environmental Protection Agency (EPA). The CWA operates through establishing water quality standards which prescribe the quality of the receiving waters[30] or through direct discharge regulations prohibiting excessive discharges of particular pollutants into waters.[31] Under the CWA a permit system is used to control direct discharges into the waters.[32] The permit from a state or federal agency would be necessary.

An additional regulatory statute is the Federal Surface Mining Control and Reclamation Act of 1977 (FSMCRA).[33] It prescribes standards for the extraction of coal on public and private lands by surface mining techniques. It sets forth several standards regulating the manner of extraction, the method of reclamation, and the operation to be applied. The Office of Surface Mining which administers the FSMCRA must assure compliance with minimum requirements concerning separation of top soil from the spoil pile, contour mining, reclamation, and compliance with other regulatory statutes.[34] The Office of Surface Mining within the Department of Interior has recently promulgated final regulations implementing the FSMCRA.[35]

Other federal regulatory programs may indirectly impact development of energy resources. Under the Resource Conservation and Recovery Act of 1976 (RCRA)[36] the EPA has promulgated proposed regulations which establish "cradle-to-grave" management of hazardous wastes. Because drilling muds and brines are considered to be "hazardous wastes," pits and other facilities storing those materials are subject to regulation and must receive permits before construction. The EPA recommends that permit applications for new pits (including those at new drilling sites) be made at least 300 days before construction.

Under the Safe Drinking Water Act the EPA is drafting a set of underground injection control (UIC) regulations.[37] These regulations will be designed to protect underground drinking water sources from injection operations. This program also requires a permitting process, which the EPA hopes to coordinate with RCRA permitting program.[38]

TRANSMISSION AND DISTRIBUTION

The impact of requirements and regulations on the transmission and distribution stage of developing a resource is probably more indirect than their impact at other stages of development. It is, however, probably a critical consideration in undertaking development of a resource. The problem specifically is land use considerations in siting transmission lines or pipelines. At both the state and federal level there are statutes regulating, for example, the location of oil and gas pipelines and requiring the permission of the state to build them. If the mineral is located at a point that is not readily accessible to transmission or distribution lines, then extraction of it may not be undertaken.

A recent illustration of this is the proposed pipeline to carry Alaskan oil from California to Texas for refining.[39] That situation also vividly illustrates the overlap, and sometimes the conflict, between state and federal policies. The use of coal slurry pipelines to move western coal to metropolitan points of use is another example of this state/federal interaction.[40]

If there is not immediate access to means of transporting the resource, then other modes of transportation must be considered. If they are going to be vehicular transportation by truck or rail both the additional cost, the energy used in the means of transportation itself, and the potential pollution through emission of pollutants by the transportation source are factors to be considered. If these violate applicable standards under the CAA, then the activity may not be permitted.

A similar problem deals with mining uranium and using it in nuclear generating plants. Those activities, including the construction and operation of nuclear plants, are all licensed and regulated closely by the Nuclear Regulatory Commission.[41] Regardless of whether the uranium ore is located on federal or private land its extraction, processing, and use is carefully controlled by the federal government.

Each of the stages of development of mineral resources discussed above also have requirements imposed for health and safety. These apply at the federal level through the Occupational Safety and Health Act (OSHA) which requires that worker safety and health be a critical factor, thus making work conditions of greater importance. OSHA regulations apply to the construction and operation of mines. They prescribe guidelines to control dust and particulate matter, for adequate ventilation, and for control of toxic pollutants.[42]

USE OF PUBLIC LANDS

At the federal level, restrictions on mineral development also exist for particular areas of public land. Land use decisions that prohibit certain kinds of activities have been made about particular areas. These activities are typified by the planning requirements of the Forest and Rangeland Resources Planning Act.[43] They also are illustrated in the Wilderness Act which has particular statutory limitations on prospecting and mining activities in designated wilderness areas.[44]

The Forest and Rangeland Resources Planning Act requires a periodic inventory of all resources on National Forest lands.[45] The Forest Service has only recently considered minerals as a separate resource element, and under regulations drafted pursuant to the National Forest Management Act,[46] it will be considering the mineral resources as part of the interpretation of geologic resource information. At this time minerals do not receive specific treatment, and no planning criteria exist to guide mineral resource development and management.

Mineral exploration and development are prohibited in most national parks and national monuments.[47] By Presidential Proclamation areas can be declared national monuments and thus removed from public lands open to mineral development. This situation recently occurred with respect to President Carter's action on some public lands in Alaska. Because in many national parks there are statutory wilderness areas, a double barrier to mineral development may exist. In areas such as the "Overthrust Belt" west of Yellowstone National Park in Wyoming, a substantial federal acreage is either restricted entirely from exploration or open only to limited exploration with no opportunity for production. In those parks and monuments where mineral and energy production has occurred, the physical disturbance and resultant visual change and stark contrast have led to public demands for an end to these kinds of activities.[48]

Under the Wilderness Act 16.6 million acres have been placed in the National Wilderness Preservation System,[49] not counting those designated for inclusion during the 95th Congress.[50] Additionally, the Bureau of Land Management is inventorying its 473 million acres for wilderness tracts,[51] the Forest Service is studying 62 million acres in RARE II,[52] Secretary Andrus has withdrawn 110 million acres in Alaska,[53] and the National Park Service and Fish and Wildlife Service are recommmending 19 million acres for wilderness.[54]

The manner of exploration or production of energy resources on public lands is also controlled under the Wilderness Act. Energy exploration in statutory wilderness areas, primitive areas, and wilderness study areas is permitted until 1983.[55] Until primitive and study areas have been nominated for statutory preservation, no leasing for production may occur. As long as wilderness study is underway or under consideration, mineral resource development on a substantial acreage is restricted.[56]

Both the Wilderness Act and the Federal Land Policy and Management Act require that assessments of mineral values be made prior to wilderness designation. Both Acts also provide for mining activities in designated wilderness areas, subject to reasonable surface protection regulations, until 1983. Assessments of mineral values have been difficult to make, however, because surface protection regulations prohibit seismic and other exploratory activities.[57]

ADMINISTRATION

The standards, requirements, and regulations are administered by a variety of agencies. At the federal level the licensing operations are undertaken principally by the agency charged with management of the public lands involved. These include the Department of Interior and the Department of Agriculture, which controls the Forest Service lands. Regulations under the Federal Surface Mining Control and Reclamation Act are promulgated and enforced by the Office of Surface Mining in the Department of Interior.[58] The standards under the Clean Air Act are promulgated and enforced by the Environmental Protection Agency.[59] Likewise, the standards under the federal Clean Water Act are developed by the EPA.[60] In the latter two instances, state programs may be promulgated which can supplant the federal program and establish a state administered permit system.[61] The licensing of nuclear activities is controlled by the Nuclear Regulatory Commission.[62]

NEPA is a unique statute in its application and administration. Its enforcement is not limited to any agency nor does it provide for standards to be promulgated in any particular manner. It applies to all federal agencies[63] and must be enforced by them when they are undertaking major federal action that significantly affects the quality of the human environment.[64] Thus its application may arise for an agency not normally involved in environmental matters[65] nor accustomed to promulgating or establishing performance standards.[66] Nonetheless its mandates must be fulfilled by the agency.

STATE REQUIREMENTS

Regulation of development of mineral resources also occurs on the state level. Typically this takes the form of regulation through a centralized commission within the state concerning a particular resource such as oil and gas. But similar requirements to the federal ones exist in many states. For example many states have adopted a "mini-EPA" type statute. The requirements of the mini-NEPA's are closely similar to the federal except that they apply to all state activities that would significantly affect the quality of the environment.[67] Thus in some instances where perhaps the federal NEPA may not apply, and EIS (typically called an Environmental Impact Report rather than statement) may be necessary anyway at the state level. Also many states have land use planning statutes for particular areas of the state. Many have statutes that are designed to protect coastal zones through coastal zone management plans.[68] Others have critical area management plans such as land use planning commissions to regulate the type developments permitted above certain altitudes,[69] in particular regions such as mountains, wetlands, or flood plains, in a state.[70] Similarly many states have laws on siting refineries, nuclear power plants, or pipelines and other activities that are ancillary to mineral development.[71] Also the state environmental quality standards must be met.[72]

An example of a state regulatory agency affecting the production of energy is the Texas Railroad Commission. The Texas Railroad Commission is a regulatory body of three elected members who oversee energy production and distribution. The Commission is empowered to regulate the drilling, spacing, and volume of production from active wells in Texas. The Commission also promulgates and enforces regulations dealing with drilling, pumping, collection, storage, and distribution of production. Many environmental issues such as the proper management of the drilling mud, drainage associated with drilling, and the disposal of drilling brine are handled by the Commission.[73]

State policies conform, often through cooperative programs, to the minimum federal standards for each area of environmental concern such as air, surface water, fish and wildlife, toxic wastes, land management, and occupational health and safety. Each state has specific administrative procedures and requirements to accomplish its legislative mandate. Most states have different boards to regulate air quality or water quality, to manage natural resources, or to conduct land use planning.

CONCLUSION

With the increased demand for energy resources and the uncertainty surrounding present sources of the resources greater attention is focusing on statutory requirements and administrative regulations that affect development of energy resources. The attention focuses not only on the substantive content of the requirements, but also on procedural aspects. The respective role of the federal and state goverment in developing resources is also being closely scrutinized. The traditional justifications of environmental protection, health and safety, preserving competition, stabilizing reasonable prices for the end product, protecting the public interest in the use of natural resources and of public lands, and providing the public input into the decisionmaking process are being re-examined. Proposals are being made to simplify and expedite the licensing process, for example, and to make less onerous performance standards and regulations concerning environmental quality. Sometimes the objectives of a proposal seem diametrically opposed to one another as in the instance of low consumer prices with maximum environmental protection.

The end result of the proposals remains to be seen. Whether exemptions are created from requirements or standards lowered will be a legislative judgment. If the past is any indicator, however, more rather than less regulation seems inevitable. And although the regulation has its impact on time and cost of developing a resource, nonetheless, its correlation to public health, safety, environmental quality, and the overall standard of living indicates a sound basis for the regulations. In that sense, the statutory requirements and administrative regulations are not intended to impede development of energy resources, but rather to insure their development in an environmentally sound manner.

NOTES

1. Speech delivered at the Fifty-fifth Annual Meeting of the Southwestern and Rocky Mountain Division, American Association for the Advancement of Science, and the Colorado-Wyoming Academy of Science, Fort Lewis College, Durango, Colorado, April 25-28, 1979.

For a fuller discussion of the topics addressed in this paper, *see* Skillern, "Environmental Law Issues in the Development of Energy Resources," *Baylor Law Review* 29:739 (1977).

2. These objectives, *inter alia,* are stated as purposes of the Department of Energy Organization Act, 42 U.S.C. § 7101, 7112.

3. See, for example, the public participation requirements in developing land and resource management plans under the National Forest Management Act of 1976, 16 U.S.C. § 1604(d).

4. For example, 42 U.S.C. § 7604 (Clean Air Act); 16 U.S.C. § 1540(c)(1) (Endangered Species Act); and 33 U.S.C. § 1365 (Clean Water Act).

5. Courts generally do not allow private citizens the right to enforce public statutes unless the legislature has clearly indicated its intent to permit private enforcement. Usually that intent must be expressly stated,but occasionally a private right to sue has been allowed based on the implied intent of Congress to permit private enforcement. The United States Supreme Court recently allowed a citizen suit for damages for alleged violations of the 1972 Civil Rights Act on the grounds of the implied intent of Congress to permit the lawsuits. Cannon v. Univ. of Chicago, No. 77-926, 47 U.S.L.W. 4549 (1979).

For a fuller discussion of the "standing to sue" issue and citizen's suit provisions, see F. F. Skillern, "Private Environmental Litigation: Some Problems and Pitfalls, 9 *St. Mary's Law Journal* 675, 684-729 (1978).

6. 42 U.S.C. § 7401 *et seq.* (1977).

7. 33 U.S.C. § 1251 *et seq.* (1977).

8. 29 U.S.C. § 655 *et seq.* (1970).

9. 42 U.S.C. § 4321 (1970) *as amended by* Pub. L. 94-83 (August 9, 1975).

10. 16 U.S.C. § 1131 *et seq.* (1964).

11. 16 U.S.C. § 1451 *et seq.* (1972) *as amended by* Pub. L. 94-370 (July 26, 1976).

12. 16 U.S.C. § 1601 *et seq.* (1976).

13. 30 U.S.C. § 181 *et seq.* (1920).

14. Ibid.

15. 29 U.S.C. § 655 *et seq.,* as amended (1970).

16. Pub. L. 94-377 (August 4, 1976).

17. For example, the Federal Surface Mining Control and Reclamation Act, 30 U.S.C. § 1201 *et seq.* (1977).

18. 16 U.S.C. § 1451 *et seq.* (1972) *as amended by* Pub. L. 93-370 (July 26, 1976).

19. 16 U.S.C. § 1131 *et seq.* (1964).

20. 16 U.S.C. § 1531 *et seq.* (1972) *as amended by* Pub. L. 95-632 (1978).

21. Tennessee Valley Authority v. Hill, U.S., 98 S. Ct. 2279 (1978) (Snail darter). Even after the 1978 Amendments to the Endangered Species Act the Tellico Dam Project that was halted in Hill was stopped by the Endangered Species Commission that was created by the Amendments. The Commission unanimously determined that it would be improper to continue the project even though it is 95 percent finished, because the project was never economical. Hence an exception to the Endangered Species Act was not granted. 9 Envir. Rptr. (BNA curr. Dev.) 1776 (Jan. 26, 1979).

22. 33 U.S.C. § 1251 *et seq.* (1977).

23. 42 U.S.C. § 7401 *et seq.* (1977).

24. Part C, Prevention of Significant Deterioration of Air Quality, 42 U.S.C. § 7470-79 (1977).

25. Visibility protection, 42 U.S.C. § 7491 (1977).

26. Ibid.

27. Part D. Plan Requirements for Nonattainment Areas, 42 U.S.C. § 7501-07 (1977).

28. Council on Environmental Quality Final Regulations, 43 Fed. Reg. 55978 (Nov. 29, 1978).

29. Ibid.

30. 33 U.S.C. § 1313 (1977).

31. Ibid. at § 1314.

32. The National Pollutant Discharge Elimination System (NPDES) is the name given the permit program established in the Clean Water Act. 33 U.S.C. § 1342 (1977).

33. 30 U.S.C. § 1201 *et seq.* (1977).

34. Ibid. at § 1251-1279.

35. The final permit program regulations were issued in 44 Fed. Reg. 14902 (March 13, 1979), to be effective April 12, 1979. The interim regulations promulgated on December 13, 1977, remain in effect until permanent program is in effect in a state. *See,* Proposed Regulations, 43 Fed. Reg. 41662 (Sept. 18, 1978). The Office of Surface Mining delayed final promulgation of the rules. 44 Fed. Reg. 1355 (Jan. 4, 1979).

36. 42 U.S.C. § 6901 *et seq.* (1976).

37. EPA Proposal for State Underground Injection Control Programs, 44 Fed. Reg. 23738 (April 20, 1979).

38. Ibid.

39. H.R. 3131, Crude Oil Transportation Amendment Act of 1979. *See* 9 Envir. Rptr. (BNA Curr. Dev.) 2135 (March 23, 1979) for discussion of the Sohio pipeline case.

40. Industry has been in an interesting position in the slurry pipeline and Sohio situations. Usually industry argues against federal regulation in preference for state control, if any. But industry advocates federal regulation to preempt the state permit process in the Sohio situation and to allow the slurry pipeline in order to avoid conflicting state/ federal regulations and differences between the state and federal governments.

41. See, Atomic Energy Act of 1954, 23 U.S.C. § 2011 *et seq.,* Uranium Mill Tailings Control Act of 1978, 42 U.S.C. § 7911 *et seq.* (1978), and Department of Energy Organization Act, 42 U.S.C. § 7101 *et seq.* (1977).

42. For a list of OSHA standards *see* 29 C.F.R. Part 1910, p. 837 (index).

43. 16 U.S.C. § 1600 *et seq.* (1974), *as amended by* the National Forest Management Act of 1976, Pub. L. 94-588 (1976).

44. 16 U.S.C. § 1133 *et seq.* (1964).

45. Note 43 *supra.*

46. Proposed rule for National Forest System Land and Resource Management Planning, 43 Fed. Reg. 39046 (Aug. 31, 1978) (first draft) and 44 Fed. Reg. 26554 (May 4, 1979) (second draft).

47. Mining Activity within National Park System Areas, 16 U.S.C. § 1901 *et seq.* (1976).

48. Hamson, D. and T. Ristau. "Death Valley National Monument—A Case Study," paper presented at Our National Landscape—A Conference on Applied Techniques for Analysis and Management of the Visual Resources, Incline Village, Nevada, April 23, 1979.

49. Council on Environmental Quality, Ninth Annual Report, p. 299 (1978).

50. See 9 Envir. Rptr. (BNA Curr. Dev.) 1183 (Oct. 20, 1978).

51. Council on Environmental Quality, Ninth Annual Report, p. 299 (1978).

52. 9 Envir. Rptr. (BNA Curr. Dev.) 1693 (Jan. 12, 1979).

53. 9 Envir. Rptr. (BNA Curr. Dev.) 1339 (Nov. 24, 1978). This action by Secretary Andrus was followed by President Carter's designating new national monuments in Alaska protecting 56 million acres. 43 Fed. Reg. 57009 (Dec. 1, 1978).

54. Ibid. at 299-302.

55. 16 U.S.C. § 1133 (1964).

56. See Parker v. United States, 309 F. Supp. 593 *aff'd* 448 F.2d 793 (10th Cir. 1971), *cert. denied,* 405 U.S. 989 (1972).

57. 36 C.F.R. § 251 and 252.

58. See note 35 *supra.*

59. 42 U.S.C. § 7101 *et seq.* (1977).

60. 33 U.S.C. § 1251 *et seq.* (1977).

61. See, 42 U.S.C. § 7410 (1977) for the use of state implementation plans under the CAA; see note 32 *supra* for the state NPDES permit system under the Clean Water Act.

62. The Energy Reorganization Act of 1974, 42 U.S.C. § 5801 *et seq.*

63. 42 U.S.C. § 4321 *et seq.* (1970).

64. Ibid. § 4332.

65. Almost every federal agency has had to prepare an impact statement on some facet of its activities.

66. The Small Business Administration and the Department of Transportation in interstate highway projects are examples of agencies not involved in establishing environmental quality standards and yet having to comply with NEPA.

67. See, for example, the California Environmental Quality Act of 1970, Cal. Pub. Res. Code § 2100 *et seq.* (West 1974).

68. See, for example, Texas Coastal Public Lands Management Act of 1973, Tex. Rev. Civ. Stat. Ann. art. 5415e-1 (Supp. 1976-1977).

69. See, for example, Vermont Land Use and Development Law, Title 10, § 6001, ch. 151, Vt. Stat. Ann. (1969) and State Land Use and Development Plans, *id.* § 6042.

70. See, for example, Florida Environmental Land and Water Management Act of 1972, ch. 72-317, Fla. Laws (1972); Massachusetts Wetlands Protection Laws, ch. 130, Mass. Gen. Laws (1965).

71. For example, California Coastal Acts of 1976, § 30,000-30,900 Cal. Pub. Res. Code and Division 21, Coastal Conservancy Act, § 31,000-31,406 (1976).

72. Most states have adopted implementation plans for the Air Act and a state NPDES permit system for the Clean Water Act as well as other state statutes providing for environmental quality protection.

73. Oil and Gas, Tex. Rev. Civ. Stat. Ann., Title 102, art. 6004-6066, specifically art. 6029a as amended.

6

INCENTIVES FOR ENERGY RESOURCE RECOVERY

Glen D. Weaver

Department of Economics
Colorado State University
Fort Collins, Colorado

SUMMARY

Hundreds of new energy development projects are planned or proposed for the Western states. Projections for Colorado alone show 77 new coal mines, 30 uranium mines, 7 oil shale facilities, 1 coal gasification plant, and an undetermined but large number of oil and gas wells. Improved cooperation between government and industry will be needed to assure proper timing and success of many of these projects. Assistance programs may include economic incentives, relaxation or clearance of environmental stipulations, streamlining of the complex permit system, increased research and development support, and expeditious leasing of federal mineral lands. This chapter examines the role of incentives in stimulating energy resource recovery, with special attention given to the synthetic fuels industry.

INTRODUCTION

The major energy problem confronting our nation just ten years ago was that of surplus production capacity. Domestic output of petroleum averaged nearly 11 million barrels per day, and an additional 1.3 million barrels of daily capacity was being withheld from the market because of the competitive advantage of importing low-cost oil from abroad (Independent Petroleum Association, 1975). Government policies then in place were designed to stimulate domestic exploration while concurrently protecting the industry against the disruptive effects of surplus supplies. This was accomplished by rewarding drilling activity with tax writeoffs, by restricting imports to a fixed percentage of internal consumption, and by prorationing domestic output to meet anticipated market demand (Mancke, 1974; Sheppard et al., 1978).

Within the past decade, our energy situation has changed dramatically. Discovery and production of both petroleum and natural gas have failed to keep pace with growing demand. The price of foreign oil has increased more than nine-fold, and our import dependence has risen from less than 30 percent to more than 40 percent (Chase Manhattan Bank, 1979; Independent Petroleum Association, 1975). Government energy policies have also changed, becoming punitive rather than supportive. The oil depletion allowance has been reduced for all companies and eliminated for some, legislative attempts have been made to force divestiture of integrated companies, price controls have been established for domestic crude oil, price controls on natural gas have been extended to include intrastate sales, and various other regulations have been imposed which discourage investment in new energy projects.

The purpose of this chapter is to examine incentives which might be used to stimulate domestic energy production. To a considerable extent, this really means the removal of existing disincentives. No attempt has been made to identify all possible options. The discussion focuses on recovery of fossil fuels, with particular attention given to the development of synthetic petroleum from oil shale. This emphasis reflects the author's greater familiarity with oil shale, together with his belief that early development of a domestic synfuels industry would enhance national security, strengthen the foreign exchange value of the dollar, and improve the U.S. bargaining position vis-a-vis OPEC's cartel pricing policies.

ECONOMIC INCENTIVES

A logical first step in promoting energy recovery would be removal of price controls on oil and gas, which are the two fuels in shortest supply. Under the recently enacted Natural Gas Policy Act, price controls on most natural gas will not be completely lifted until 1985 or 1987, and controls on some gas will remain indefinitely (Federal Energy Regulatory Commission, 1978). In April of 1979, President Carter announced that he will phase out oil price controls by September 1981. He also recommended that Congress enact a windfall profits tax that would divert about 13 percent of the industry's increased earnings to an "energy security fund" to be used for mass transit improvements, research on alternative energy sources, and aid to families hard-pressed by rising fuel bills.

The rationale of the President's tax proposal is debatable. In any event, if the objective is to stimulate exploration and development now, then price controls on both oil and natural gas should be lifted immediately. Phasing in of the market price will clearly, to a greater or lesser degree, encourage operators to postpone drilling or development until the highest price can be obtained. Phased decontrol will also create confusion and frustration within the industry as operators attempt to comply with the regulations and obtain their perceived equitable share.

Incentives other than price decontrol will be needed to commercialize the liquefaction of oil shale. Although the physical resource base of this fuel is extraordinarily large, so too is the financial risk of bringing it onstream (Energy Development Consultants, 1979). Capital and operating costs will undoubtedly be high, but just how high is a matter of some speculation. The most advanced technology for shale oil recovery, which involves mining and surface retorting of large tonnages of rock, has been tested only at the semiworks level. Commercialization would require scaling up by a factor of ten. Equipment which has functioned well in the semiworks plants may require expensive design changes to be operable at large scale. Developers of the modified *in-situ* process can testify to the problems associated with technological scale-up. Their experimental commercial-sized chimneys have recovered only 20 to 40 percent of the inplace shale oil rather than the 60 percent or more that was anticipated (Gill, 1978).

Direct economic incentives to the synfuels industry could take a variety of forms, including accelerated depreciation of capital costs, loan guarantees, convertible grants, price supports, or tax credits on product output. To date, all efforts to win Congressional approval of incentive programs have been singularly unsuccessful. The latest proposal, a $3 per barrel tax credit on shale oil, was defeated in 1978 but was reintroduced during the 1979 session. A tax credit has the advantage of posing no significant loss to the federal treasury if shale oil ventures prove unsuccessful, since payments would be made only on the basis of actual production. As proposed in H.R. 1969, producers would receive a $3 per barrel credit until the reference price, defined as the landed price of Arabian marker crude adjusted for inflation, reached $20.50. Beyond this value the credit would be incrementally reduced and finally terminated when the reference price reached $23 per barrel. With marker crude selling in the U.S. for about $16.50 per barrel in early 1979, and with production costs for raw shale oil in the range of $15 to $27, it appeared that the tax credit proposal might stimulate some companies to move forward on their commercialization plans.

Producers of conventional petroleum have long been given special tax benefits. One of the existing writeoffs, the foreign oil tax credit, allows companies to classify production payments to foreign governments as income taxes instead of royalties, thereby giving them a higher deduction on their U.S. income taxes (Sheppard et al., 1978). This credit would seem to encourage investment abroad at the expense of domestic operations. In 1977, the Chase Manhattan group of petroleum companies, most of whom are U.S.-based, allocated 44 percent of their exploration-production expenditures to overseas operations (Chase Manhattan Bank, 1978). Elimination of the foreign tax credit would encourage U.S. companies to focus more strongly on domestic projects.

RESEARCH AND DEVELOPMENT SUPPORT

A second alternative for stimulating energy resource recovery is to expand research and development support, including funding for commercial demonstration projects. Table 6.1 shows the percentage allocation of energy-related research and development (R&D) expenditures over the past four years, together with projected outlays for the upcoming fiscal year. Note that direct outlays for nuclear fission and fusion programs have comprised about 40 percent of total outlays. If related expenditures in the supporting sciences are included, the proportion rises to at least 44 percent. In contrast, R&D support for fossil, solar, and geothermal technology has collectively averaged only 30 percent of total outlays. The efficacy of the

Table 6.1. Department of Energy civilian R&D outlays, fiscal years 1976—1980

	Percentage Allocation					
	1976	1977	1978	1979	1980	1976—79
Nuclear fission	35.1	37.1	31.1	28.7	24.8	30.6
Nuclear fusion	11.2	11.6	8.4	9.1	9.5	9.8
Fossil energy	16.7	16.0	18.6	21.8	20.0	19.0
Oil shale	(0.5)	(0.5)	(0.7)	(1.3)	(0.8)	(0.8)
Solar energy	4.3	6.6	9.8	8.5	11.3	8.6
Geothermal energy	1.4	1.8	2.5	3.6	3.5	2.7
Conservation	2.8	4.3	7.2	6.5	7.0	5.9
Environmental research	9.8	6.3	6.1	5.4	5.8	6.4
Basic sciences	8.9	11.4	11.0	11.1	11.7	11.0
Basic energy sciences	9.4	4.9	4.9	5.3	6.4	5.9
Other	0.4	—	0.2	—	—	0.1
Billions of dollars	2.0	2.8	3.3	3.7	3.8	3.1

Source: Chemical and Engineering News (1976—79)

current R&D mix can be questioned, particularly in view of the uncertain commercial future of nuclear fission.

Total outlays for the upcoming fiscal year are projected to increase by only 2.5 percent over 1979 expenditures, with support for oil shale projected to actually decline by 40 percent. According to the President's Office of Science and Technology, this reduced spending reflects the view that federal R&D programs over the past five years have successfully advanced near-term technologies, and, with increases in energy prices and other incentives for private investment, less reliance should be placed on the federal budget to meet national needs.

As applied to oil shale conversion, this view is correct in the sense that basic research and development have reached an advanced stage. Hence the most effective policy now would be to provide financial support for the commercial demonstration phase. In 1978, Congress authorized the Department of Energy to share up to 75 percent of the cost of an aboveground retorting module, and $15 million in start-up funds were appropriated for the project. The Department has recently been soliciting proposals from interested private firms. Total cost of the module program is expected to range between $150 and $250 million.

IMPROVED ACCESS TO FEDERAL LANDS

A third incentive option is that of improving access to federal mineral lands. The federal government holds title to approximately one-third of the nation's total land area, with ownership in the eleven Western states averaging nearly 50 percent (U.S. Bureau Land Management, 1977). Federal control also extends over the outer continental shelf, and in many inland areas the government has retained mineral rights to lands that are now privately owned. For example, federal agencies control one-third of the surface acreage in Colorado but two-thirds of the mineral rights (U.S. Bureau of Mines, 1978, Table 4).

Public lands are important target areas for oil and gas exploration, uranium discoveries, and geothermal resource recovery. About 80 percent of the richer oil shale deposits in the Rocky Mountain states and 72 percent of the coal reserves in the Rocky Mountain-Northern Great Plains states are federally owned (U.S. Department of Interior, 1973, Vol. I; U.S. Department of Interior, 1979, p. 2-5).

Accessibility to federal lands under the Mineral Leasing Act of 1920 has been substantially reduced within the past decade. Independent studies by the Department of the Interior and the Congressional Office of Technology Assessment show that nearly 40 percent of the onshore acreage was formally closed to exploration and development in the mid-1970s, with access to another 24 to 29 percent being either moderately or severely restricted (Task Force, 1977, Table 9). The inventories did not include all administrative actions that restrict mineral development, such as the Department of the Interior's coal leasing moratorium which has been in effect since 1971.

Some parties may argue that the access problem is not very serious because many of the restrictions are temporary, examples being those lands affected by Forest Service and BLM wilderness proposals or lands withdrawn under the Alaska Native Claims Settlement Act. However, such "temporary" withdrawals can last for several years or, in the case of Alaskan lands, perhaps for decades. Also, the restoration of lands temporarily withdrawn will be at least partially offset in the future by growing public demand for creation of new parks, endangered species habitat, and similar wilderness-type areas.

IMPROVING THE PERMIT SYSTEM

The last incentive option to be discussed is that of streamlining and stabilizing the regulatory permit system. Currently, energy developers must thread their way through a bureaucratic maze of paperwork in order to secure the permits required for exploration, development, and production activities (Davidson, 1976; Novak, 1976; Office of Technology Assessment, 1979). Environmental stipulations contained in the permits are subject to change on short notice, and entirely new regulations may be imposed by either legislative or administrative action. The problem is especially acute for developments involving federal lands because, in addition to incurring more government regulation, they may also be harrassed by citizen lawsuits contesting implementation of the National Environmental Policy Act.

The complexity of the existing permit system is illustrated in Table 6.2, which shows the bureaucratic paperwork system confronting the oil shale industry in Colorado. Some of the governmental agencies merely stipulate regulations that must be complied with. Others issue both regulations and permits. Still others give review and clearance to a permit application required by another agency. For example, the National Park Service and State Historical Society oversee preservation of archaeological and historical sites. Their clearance must be obtained before the lead agency will issue its permit.

The bureaucratic paperwork adds directly to the cost of energy development projects because additional personnel must be hired. It also provides opportunity for disruption of company planning schedules,

Table 6.2. Regulators of oil shale projects in the State of Colorado

	Mines & Process	Power Plants	Transmission	Pipelines	Railroads	Roads
FEDERAL GOVERNMENT						
Department of Interior						
Bureau of Indian Affairs	P	P	P	P	P	P
U.S. Geological Survey	P	C	C	C	C	C
Mining Enforcement & Safety	R					
Bureau of Land Management	P	P	P	P	P	P
Bureau of Reclamation	P	P	P	P	P	P
National Park Service	C	C	P	P	P	P
Fish & Wildlife Service	C	C	C	C	C	C
Department of Agriculture						
Forest Service	P	—	P	P	P	P
Soil Conservation Service	C	C	C	C	C	C
Rural Electric Administration	—	P	P	—	—	—
Environmental Protection Agency						
Water quality	P	P	R	R	R	R
Air quality	P	P	—	—	—	—
Solid waste disposal	R	R	—	—	—	—
Hazardous materials	R	R	—	R	—	—
Department of the Army						
Corps of Engineers	P	P	P	P	P	P
Department of Labor						
Occupational Safety & Health	R	R	R	R	R	R
Department of Transportation						
Federal Aviation Administration	P	P	P	—	—	—
Federal Highway Administration	C	C	C	C	C	C
Materials Transportation Bureau	R	—	—	R	R	—
Federal Power Commission	—	P	P	—	—	—
Federal Communications Commission	P	P	P	P	P	P
Interstate Commerce Commission	—	—	—	P	P	P
Nuclear Regulatory Commission	P	P	—	—	—	—
STATE OF COLORADO						
Department of Health						
Air Pollution Control Division	P	P	—	—	—	—
Water Pollution Control Division	P	P	—	—	—	—
Engineering & Sanitation Division	P	P	—	—	—	—
State Historical Society	C	C	C	C	C	C
Department of Highways						
Division of Highways	P	P	P	P	P	P
Highway Safety Division	R	R	R	R	R	R
Department of Labor						
Co. Occupational Safety & Health	R	R	R	R	R	R
Department of Natural Resources						
Div'n of Water Resources (State Eng)	P	P	—	—	—	—
Geological Survey	C	C	—	—	—	—
Co. Groundwater Commission	P	P	—	—	—	—
State Board of Land Commissioners	P	P	P	P	P	P
Division of Mines	R	—	—	—	—	—
Oil & Gas Conservation Commission	P	—	—	—	—	—
Mined Land Reclamation Section	P	—	—	—	—	—
Co. Soil Conservation Board	C	C	C	C	C	C
Co. Water Conservation Board	C	C	C	C	C	C
Division of Wildlife	C	C	C	C	C	C
Division of Parks & Recreation	C	C	C	C	C	C
Public Utilities Commission	—	P	P	P	P	—
Land Use Commission	C	C	C	C	C	C
LOCAL GOVERNMENTS						
County						
Land use	P	P	P	P	P	P
Air quality	P	P	—	—	—	—
Water quality	P	P	—	—	—	—
Health	R	R	—	—	—	—
Fire	R	R	R	R	R	R
Flood	R	R	R	R	R	R
Building codes	P	P	—	—	—	—
Roads	R	R	P	P	P	P
Municipal & special districts						
Land use	P	P	P	P	P	P
Air quality	P	P	—	—	—	—
Water	P	P	—	—	—	—
Sanitation	R	R	—	—	—	—
Health	R	R	—	—	—	—
Fire	R	R	R	R	R	R
Flood	R	R	R	R	R	R
Building codes	P	P	—	—	—	—
Streets	R	R	P	P	P	P

Legend: P = Permits; R = Regulations; C = Clearance or Review contributes to inflationary costs when delays do occur, promotes additional costs if designs have to be changed, and poses the uncertainty that an insurmountable roadblock will eventually be encountered. All permit applications cannot be filed when work on the project first commences. Instead, they must track with site acquisition progress and with design and engineering progress throughout the 5 to 10 years required for completion of a major project. Agencies receiving the applications work within their own time frame of review, which may not coincide with the planning schedule of the development company. Disruptions are especially likely to occur at those points in the permit process where the public can make comments or special interest groups can intervene. If contested issues cannot be resolved satisfactorily, then the agency has the option to deny the permit. Even after the permit is issued, there remains the possibility that environmental stipulations will be altered at a later date.

Problems associated with the permit system adversely affect small as well as large companies. What has historically been an advantage of the smaller companies in flexibility of rapid operation has now become a penalty because of delays requiring longer term commitments on capital and payments for down-time when personnel and equipment cannot be used. A survey conducted in 1977 revealed that some small operators had either ceased or were substantially decreasing their oil and gas drilling on federal lands because of the extra cost and delay that accompanies work on federal leases (Everett, 1977).

Several actions could be taken to streamline and stabilize the permit process. One approach would be to provide a clearinghouse for permit applications. Currently, energy developers expend considerable time just determining which regulations apply to which of their proposed actions and which agency has jurisdiction. One of the Colorado counties has responded to this problem by placing all of its energy-resource facility permits within the local planning office. Separate applications must still be filed for different permits, but the developer only has to deal with one set of people, all in one place.

Another step would be to minimize duplication of regulations between federal, state, and local governments, and to eliminate interpretative differences between federal and state policies that implement the same federal legislation. To this end, efforts of the Environmental Protection Agency to turn over to the states responsibility for enforcing federal regulations should be accelerated.

In some cases, it may be desirable to enact legislation that would exempt developers from any changes in environmental regulations specified in their permits. Should the national energy situation continue to worsen, it might even become desirable to designate selected parts of the country as "energy reserves," or areas within which energy projects could proceed under less stringent environmental regulations than presently exist. Candidates for energy reserve status in the Western states would include the oil shale lands in the Piceance-Uinta basins, the prospective oil and gas areas in the Overthrust Belt (see Chapter 1 by Dresher), and the coal lands in the Powder River Basin.

CONCLUSION

In conclusion, energy resource recovery could be accelerated by providing direct economic incentives, by expanding research and development support, by improving access to federal mineral lands, and by streamlining and stabilizing the regulatory permit system. The most effective incentive program would be a coordinated one that integrates all appropriate options. Decontrol of oil and gas prices, for example, will not be fully effective if companies are denied access to drilling prospects on federal lands. Nor will a tax credit on shale oil stimulate much synthetic fuels production if development is stymied by permit delays or by the imposition of new environmental regulations that increase costs and limit ultimate production levels. In other words, we are still in need of a comprehensive national energy policy.

REFERENCES

Chase Manhattan Bank. 1978. *Financial Analysis of a Group of Petroleum Companies.* Chase Manhattan Bank, New York, 32 pp.

Chase Manhattan Bank. 1979. "United States Petroleum Highlights," *The Petroleum Situation,* 3(3): 4.

Davidson, D. 1976. "The Permit Procedure: Many Steps, Many Stops," *Shale Country,* 2(11): 5-6.

Energy Development Consultants, Inc. 1979. *Oil Shale in Colorado 1979.* Colorado Energy Research Institute, Golden, Colorado, 90 pp.

Everett, A. G. 1977. "Remarks to the API Committee on Public Issues Concerning the Status of the Public Lands Project." Houston, Texas, unpublished.

Federal Energy Regulatory Commission. 1978. "Natural Gas Policy Act of 1978 Fact Sheet." Federal Energy Regulatory Commission, Washington, D.C., 11 pp.

Gill, D. 1978. "Shale Production, Test Projects on the Increase But Obstacles Still Abound," *Western Oil Reporter,* 35(8): 61-65.

Independent Petroleum Association of America. 1975. "United States Petroleum Statistics." Independent Petroleum Association of America, Washington, D.C.

Mancke, R. B. 1974. *The Failure of U.S. Energy Policy.* Columbia University Press, New York, 189 pp.

Novak, A. 1976. "The Shale Paperwork Maze of Permits, Clearances, Regulations," *Shale Country*, 2(11): 4-5.

Office of Technology Assessment. 1979. *Analysis of Laws Governing Access Across Federal Lands: Options for Access in Alaska*. Govt. Printing Office, Washington, D.C., 260 pp.

Sheppard, W. J., B. Gordon, S. Solomon, D. D. Moore, and J. S. Fattorini, Jr. 1978. "Oil Energy Incentives," in Battelle Pacific Northwest Laboratories, *An Analysis of Federal Incentives Used to Stimulate Energy Production*. Government Printing Office, Washington, D.C., pp. 201-235.

Task Force on the Availability of Federally Owned Mineral Lands. 1977. *Final Report of the Task Force on the Availability of Federally Owned Mineral Lands*. Govt. Printing Office, Washington, D.C., 103 pp.

U.S. Bureau of Land Management. 1977. *Public Land Statistics 1976*. Govt. Printing Office, Washington, D.C.

U.S. Bureau of Mines. 1978. "Minerals in the Economy of Colorado," *State Mineral Profile 30*. U.S. Bureau of Mines Branch of Distributions, Pittsburgh, Pennsylvania, 17 pp.

U.S. Department of the Interior. 1973. *Final Environmental Statement for the Prototype Oil Shale Leasing Program*. Govt. Printing Office, Washington, D.C., V. I.

U.S. Department of the Interior. 1979. *Final Environmental Statement, Federal Coal Management Program*. Govt. Printing Office, Washington, D.C.

7

PROCESSES FOR ENERGY RESOURCE RECOVERY IN DESERT REGIONS

Roshan B. Bhappu

Mountain States Research & Development
Tucson, Arizona

SUMMARY

Like any other part of the earth, nature has endowed the arid and semiarid regions with valuable mineral resources including energy resources such as coal, oil shale and uranium. However, the difficulty of obtaining an adequate water supply at reasonable cost has hampered the development of these resources in water-deficient areas. This chapter is concerned with the dry and low-water consuming mineral processing and extractive technologies currently available for recovering these energy mineral resources. The chapter also covers potential extractive processes and techniques that could be used in the future and provides guidelines for research and development input necessary for extracting these resources.

INTRODUCTION

Man, by his ability to utilize the mineral resources of the earth, has made tremendous progress in a relatively short geological time. At the dawn of civilization, he used crude stones, flint, chert, and bone implements which were the only useful raw materials available. Chaldeans, Babylonians, and Egyptians made extensive use of clay for pottery, bricks, and tiles. Neolithic man was acquainted with native metals such as gold, copper, and tin, mostly from placer sources. Herodotus (484?-425 B.C.), Theophratus (372-287 B.C.), Strabo (A.D. 19), Avicenna (980-1037), and

"

Agricola (1494-1555) in his famous *De Re Metallica* (see 1950 translation), have all described the occurrences, methods of extraction, and uses of metals, gemstones, earths, and salts. At the termination of the Dark Ages, the major metals in use were iron, copper, lead, tin, gold, silver, and mercury.

With the advent of the Industrial Revolution three centuries ago, man learned to produce iron from ores using coal and mechanical power. Since that time the use of metals increased rapidly and each progressive step in the history of man brought about increasing demand for old metals and the discovery and development of newer ones. During World Wars I and II, more metals and minerals were used than in all preceding history.

Today more than one hundred minerals are involved in international trade, some being necessary and vital for our essential industries, while others contribute to nonessential industries or luxuries. Regardless of the end use, it is evident, that with increasing population and with the industrialization of developing countries, the demand for our mineral resources will continue to increase.

Like any other part of the earth, nature has endowed the arid and semiarid areas with valuable mineral resources. However, the difficulty of obtaining an adequate supply of water at reasonable cost is one of the major problems seriously hindering the economic development of these resources in such water-short regions. This problem is particularly acute in regard to the extraction of mineral values from naturally occurring ores using the well-established mineral processing and extraction processes which require considerable amounts of water. For this reason, the use of dry concentration and extraction techniques requiring no or limited amounts of water should be considered for the development of mineral resources in those areas.

In the United States a major portion of our energy mineral resources such as coal, oil shale and uranium are located in the arid regions of the West and Southwest. The chapter by Dresher on "Energy Resources Located in Arid Lands" (Chapter 1) clearly illustrates the location of these resources. Most of the uranium and an appreciable portion of the total coal produced in the United States comes from this region. Moreover, the major portion of the oil shale deposits in the nation is also located in this arid region. Because of the depletion of our oil and gas resources it is inevitable that the above energy resources located in the desert regions of the United States will become increasingly important. Accordingly, it is imperative that we develop extraction techniques which will produce the required fuels effectively from the above resources located in the arid regions.

This chapter is concerned with the extraction techniques available for the processing of coal, oil shale and uranium located in water-deficient regions of the United States. These include conventional mining and processing of coal using dry beneficiation and low-water consuming concentration techniques; *in situ* gasification of coal; conventional mining and processing of oil shale as well as *in situ* processing of shale; and *in situ* extraction of uranium. The chapter also includes consideration of environmental problems associated with the specified techniques. Finally, the chapter suggests new research areas for the successful development of commercial processes for treating the fuel minerals located in the arid regions of the United States.

EXTRACTION OF ENERGY MINERAL RESOURCES IN ARID REGIONS

Exploration

Because of the random distribution of mineral resources throughout the world, it is likely that as many mineral deposits including energy mineral deposits are located in arid regions as there are in the wet areas. Such deposits are easy to locate and to identify due to lack of vegetative cover and have been mined extensively. They are currently the principal economic element in large parts of the arid regions and these resources may be expected to assume even greater significance in the future (Cloud 1969).

In the current exploration activities, geologists and geophysicists are using modern airborne equipment for reconnaissance exploration equally effectively over jungle and deserts. In recent years, the increase in success factor in the search for hidden mineral deposits can be directly attributed to the dynamic evolution of exploration geophysics as a spin-off resulting from significant contributions in the field of space geo-sciences; especially, due to the development of super-sensitive detecting, measuring, and recording equipment which allow the pickup of weaker and weaker signals against backgrounds of geologic and instrumental noise. Such techniques are specifically effective in desert regions covered with sand dunes and alluvium.

Mining

After a mineral deposit has been located, explored, and its ore reserves and grades determined, the next phase is the recovery of valuable metals or minerals by mining. Over the years mining has evolved into a very scientific and exact engineering discipline and, currently, a large number of surface and underground mining methods are available for mining any type of mineral deposit even in the arid and semiarid regions as long as the economics are favorable. Thus, today there are surface coal mines capable of mining 30,000 to 100,000 tons per day of coal and waste and there are also underground mines capable of mining relatively thin coal seams as deep as 1,000 to 2,000 feet in Kentucky and Pennsylvania. Moreover, new mining techniques such as *in situ* gasification of coal, solution mining of uranium and ocean mining for manganese nodules are now being developed to extract mineral resources heretofore not feasible.

Metallurgy

Metallurgy may be defined as the art and science of extracting metals and minerals economically from their ores and other metal-bearing products and adapting these for human utilization. The field of metallurgy includes all the concentration processes and techniques involved in separating the valuable metals or minerals from the worthless host rock (gangue) as well as the smelting, refining, and working of the extracted minerals. In achieving this goal, metallurgists utilize all the basic laws of physics, chemistry, and even biology to the concentration, extraction, purification, alloying and working of metals to meet the demand of our minerals-based society. Figure 7.1 lists the various processes involved in extractive metallurgy and shows their interrelations (Bray 1947).

Mineral Processing

Mineral concentration, mineral processing or ore dressing is the art and science of treating natural ores to separate valuable minerals (concentrates) from worthless gangue (tailings) on the basis of physical or chemical properties of the minerals. The concentrate is next subjected to a chemical treatment to break up the metal-bearing compounds to liberate the metallic elements. Such chemical processes are quite expensive as a rule, and the cost is proportional to the bulk of material treated. Thus, it is economically more advantageous to treat a small amount of enriched concentrate than a large volume of low grade ore.

Separation of valuable minerals on the basis of physical properties may depend on the size, shape, and color of the particles; on their magnetic susceptibility; and on their electric or heat conductivity. The above separations can be carried out in water or in air under dry conditions. In

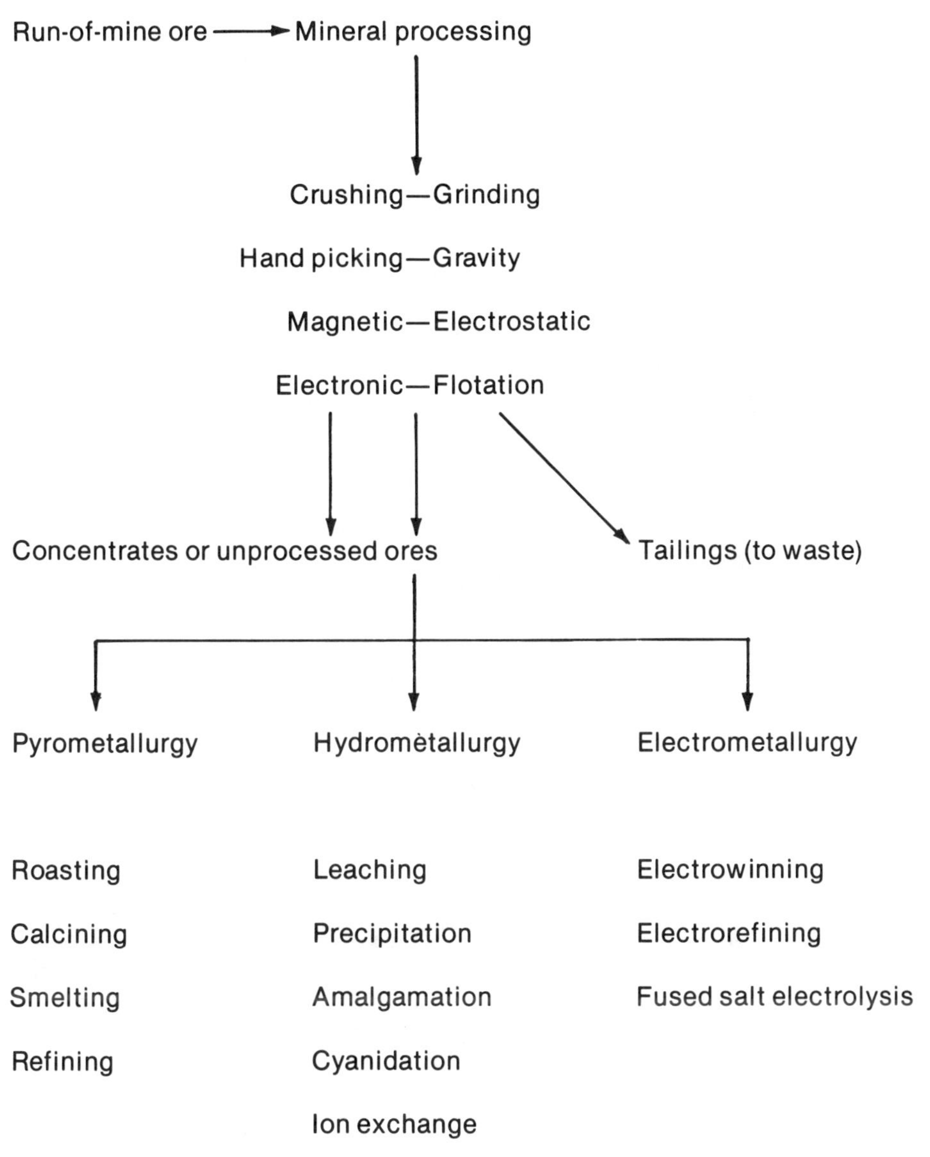

Figure 7.1 Process involved in extractive metallurgy

general, wet processing shows a more effective separation than dry concentration.

In mineral processing the amount of new water required depends on the method employed and on the extent of water reclamation utilized for recovering process water for recycling. Table 7.1 lists the water-to-ore requirements for the known concentration methods (Deju, Bhappu, Evans, and Baez 1972).

As can be seen, the majority of the wet concentration methods require a rate of 2:1 to as much as 20:1 for water-to-ore treated. However, it is also clear that with appropriate water reclamation systems it is possible to recover 75 to 90 percent of the process water for recycling. Thus, even in water-deficient areas wet processing cannot be ruled out for the treatment of a given ore. It may be that the underground mine itself will provide the required water for processing.

Assuming that there is very little water available for processing, there are several dry processing methods and techniques available for recovering energy mineral resources in water-deficient areas. These processes include (United Nations 1968):

1. Dry concentration
2. Differential disintegration
3. Magnetic separation
4. Electrostatic separation
5. Electronic separation
6. Radio-activity sorting
7. Heavy media separation
8. Thermal concentration

The above-mentioned methods clearly indicate that the majority of dry concentration procedures are based on physical properties of the mineral such as separating lighter coal from heavier refuse by dry pneumatic tabling, or separation of pyrite from coal by high gradient magnetic separation. To a lesser extent, techniques are also available for separation,

Table 7.1. Water requirements for various concentration methods

Process	Water : Ore	Water reclaimed
Scrubbing and washing	4-20 : 1	up to 50%
Gravity concentration	4-15 : 1	up to 75%
Flotation	2-4 : 1	up to 80%
Hydrometallurgical	2-1 : 1	up to 85%
Heavy media separation	2-1 : 1	up to 90%
Dry concentration	0 : 1	—

based on the chemical properties of the minerals such as dissolution of uranium minerals in acid or carbonate-bicarbonate solutions.

While some dry concentration methods such as gravity, thermal concentration, and magnetic separation have been practical for years, other techniques, such as electrostatic separation, sorting by electronic selection and heavy-liquid separation have resulted from very recent technological advancements. Through the application of such newer procedures, a large number of metallic, non-metallic and energy minerals can now be recovered with a minimum use of water at or near the mine.

Dry Gravity Concentration. It is interesting to note that most of the valuable minerals of base and rare metals have higher specific gravities (with specific gravity ranging from 3.0 to 19.0). Accordingly, the separation of these valuable minerals from their ores is relatively easy, such as separation of uraninite (sp.gr. = 10.7) from quartz (sp.gr. = 2.6) or separation of coal (sp.gr. = 1.4) from refuse (sp.gr. = 2.6). On the other hand, if two or more minerals with similar specific gravities are present in the ore (such as oil shale and slate, both with specific gravities around 2.5) the separation of one from the other would be difficult.

Currently, a large number of dry (pneumatic) concentration devices are available for economic recovery of valuable minerals in the arid areas. These devices use air to effect differential movement of mineral particles of varying specific gravities, shapes and sizes. These include pneumatic jigs and tables, fluidized bed separators using fine particles of a heavy substance in air having an apparent density intermediate with the mineral species to be separated, and air classifiers and cyclones utilizing centrifugal action.

Typical minerals concentrated by dry gravity separation devices include coal, uraninite, uranothorianite and monazite.

Differential Disintegration. These methods have been successfully used for dry concentration of various minerals such as clays, micas, cement rock, barite, iron ores, carnotite, asbestos, spodumene, fluorite, coal and many other minerals. The techniques involved in differential disintegration are either based on differences in hardness of minerals (scrubbing, scuffing, and differential grinding) or on internal stresses set up by heat (decrepitation).

The equipment involved for differential dry grinding, scrubbing and scuffing include attrition mills, impact mills, jet or fluid energy mills, pan or chilean mills, rolls and roller mills, and tumbling mills. In recent years, the development of the Aerofall mill (autogenous or semi-autogenous mills) and the Majac mill (jet pulverizer) systems have been used increasingly for differential grinding and concentration of selected minerals including those of uranium.

Magnetic Separation. Magnetic separation of minerals in a dry state uses differences in their magnetic attractability. Besides separation of iron ores, magnetic separation can be effectively used for dry concentration of uranium minerals such as pitchblende. Currently, a large number and types of very sophisticated magnetic separators are available.

Electrostatic Separation. Electrostatic separation, on the other hand, is based on differences in electric charging by fraction (tribo effect), electrification by heating (pyroelectric effect), dielectric constants, and conductivities of different minerals. In addition, separation may also be based on ionization and electric wind attending corona discharge in gases (corona method). In recent years, considerable research and development in this field has taken place in the USSR and based on their efforts, a large nubmer of metallic and industrial minerals as well as energy minerals can be effectively separated in a dry state. Classical examples of electric methods of concentration of energy minerals include coal fines, oil shale and uranium containing carbonaceous minerals.

Electronic Separation. Electronic separation of minerals utilizes sensitive instrumentation to detect differences in radiation emission, reflectivity, color and luster of minerals. The elements necessary for effective mechanical electronic sorting devices are: (a) presentation of mineral pieces for examination by sensors: (b) a sensing device for selecting the pieces to be removed; (c) an electronic system to act upon the information provided by the sensing device; and (d) a means of removing the selected pieces.

At the present time, a large number of commercial electronic separators are available for effective separation of various minerals. These include Lapointe Picker (radio-activity sensed by beta-sensitive Geiger tube) for uranium ore processing; Goodman Sorter (X-ray detector) for molybdenite and tin ores; Sortex Separator (based on reflection or transmittance of light) for separating calcite from dolomite; Gromax Electronic Sorter (based on light reflectance, color and shade differences) for barite, talc, and titanium ores; and Ore Sorters (based on electrical conductance and reflected or transmitted light) for processing iron ore.

It should be noted that some of these concentration methods are still in their infancy and that there are possibilities of using many other sensing properties and devices. These include electromagnetic spectrum, infrared rays, chemical staining of selected minerals, fluorescence, and neutron activation.

Heavy Liquid Separation. Heavy liquid separation utilizes organic liquids of selected specific gravities to obtain separation between the valuable and the gangue minerals. Thus by using tetrabromo-ethane of specific gravity 2.94, the typical gangue minerals (silica and silicates) with specific gravity of 2.6 to 2.8 can be separated from valuable minerals (oxides and sulfides) having specific gravities above 2.94. In this case, the gangue minerals sink in the liquid thereby concentrating the valuable minerals. Since no water is required in this separation process, this technique can be classified as a dry concentration process.

In the past, this method was not widely used because of the lack of inexpensive industrial heavy liquids of sufficiently high specific gravities. Today, heavy organic liquids formulated with a bromine base are available in tonnage quantities at economically favorable prices. In recent years, considerable research and development has been carried out in Israel on this versatile process. Besides tetrabromo-ethane (2.94 sp.gr.), the other commercially available liquids include: methylene iodide (3.31 sp.gr.), bromoform (2.89 sp.gr.), and methylene bromide (2.49 sp.gr.). The industrial applications of heavy liquid separation for selective concentration of valuable minerals include: coal, shale and some uranium minerals.

Thermal Concentration. Thermal conentration is one of the most widely used methods of processing ores to recover valuable minerals, compounds and metals. In this process, minerals are separated by the application of heat. In many metal extraction practices this would be a preliminary step to an actual reduction (pyrometallurgical) process.

It should be noted that most of the arid and semiarid areas of the world contain natural fuels (oil, gas, and coal) which can be readily used for thermal concentration of ores and pyrometallurgical recovery of metals. A classical example would be the extraction of oil from oil shale by retorting in a surface facility or retorting under an *in situ* environment.

EXTRACTION TECHNOLOGY

Once the desirable minerals are concentrated, the next steps in extraction systems are: (a) to decompose the mineral to free the metallic component as a crude product and (b) to refine this impure product into a marketable or commercial grade commodity. Since this phase of extraction involves chemical properties of mineral rather than physical (as is generally the case in concentration processes), it is often referred to as "chemical processing" whose three important subdivisions are pyrometallurgy, hydrometallurgy, and electro-metallurgy.

Ever since the first metal was extracted from an ore by the action of wood fire in ancient times, the pyro or fire metallurgy procedures have played a significant role in the processing of various minerals. A large number of major metals such as iron, nickel, tin, copper, zinc, lead, gold, and silver as well as many minor and rarer metals such as molybdenum, tungsten, and titanium are won from their ores or concentrates by pyrometallurgical methods. Treatment of oil shale, too, involves basic principles of pyrometallurgy.

Smelting is a major pyrometallurgical process in which the high grade ore or concentrate is mixed with suitable fluxing compounds and reducing agent such as carbon or fuel oil and reacted in appropriate furnaces to produce an impure metal and a worthless slag. The slag, being lighter than molten metal, floats on top of the molten mass and is readily removed. The crude metal is next refined by either heating in the presence of purifying reagents or by electrorefining to produce pure marketable products.

For those materials which do not readily respond to physical concentration of valuable minerals from gangue, suitable chemical or hydrometallurgical processing is applied for extracting the metal into a salable product. In such procedures, the ores or concentrates are digested with selective solvents to dissolve the valuable metals or minerals and then the metal bearing solutions are treated further by concentration, precipitation or electrolysis to recover the metals or compounds. In recent years, the advent of selective leaching reagents (lixiviants), specific extraction techniques (solvent extraction and ion exchange), and the use of autoclaves (using high pressures and temperatures) have allowed metallurgists to recover metals and compounds more effectively and economically from refractory and lower grade resources. Nearly all of the uranium ores are processed by hydrometallurgical means to produce yellow cake.

Of course, in arid and semiarid regions, due to lack of water, hydrometallurgical processes for the recovery of uranium may not appear attractive. Nevertheless, as demonstrated by uranium processing plants in the Southwest, under favorable conditions there is a place for hydrometallurgical extraction in desert areas, especially since it is possible to recover and recycle 70 to 75 percent of the process water. Also, it is entirely possible that in certain cases, the underground mine itself will provide the required water for processing.

It should also be noted that, in general, process water need not be pure for it is entirely feasible to use brackish water for processing. Especially, physical concentration of minerals in a wet state is not affected by the quality of water. Furthermore, certain flotation processes involving selective recovery of sulfide minerals does not require pure water and separations can be readily achieved in brackish water or sea water. Even in hydrometallurgical extractions, presence of chloride ions under acid leach conditions may increase the dissolution rate of metals and minerals. However, the most important detrimental concern with the use of brackish or sea water for processing is its corrosive nature, abatement of which requires the use of expensive materials of construction and constant maintenance of pumps and pipelines.

Since the art or science of chemical processing deals in the separation of valuable metals from gangue materials either in the solid, liquid or gaseous phase, it can be readily appreciated that mineral processing engineers and scientists have a special competence to understand and to remedy pollution problems associated with metal extraction. The same basic principles that govern chemical processing also control pollution abatement technology and most of the common processing machinery can be applicable for solving pollution problems. Accordingly, it is up to the mineral engineers and metallurgists to pave the way for improvements and innovations in pollution abatement technology and to share their knowledge with environmental engineers in providing the badly needed mineral resources and metals with a minimum degradation of the environment.

COAL

Coal mining, preparation, transportation, and utilization have undergone considerable technical innovations since World War II. Significant advancements in mechanization of coal mines have resulted in increased production with lower operating costs. Modern coal mining is characterized by the advent of continuous mining machines, in mechanical loading of coal underground and by significant improvement in strip mining due to new types of earth-moving equipment.

Along with advances in mining methods and machinery, several more efficient techniques of coal cleaning and preparation have been developed. These include cleaning of fine coal fractions by heavy media separation techniques and by selective flotation of finer particles.

One of the major problems facing the coal producers in the Southwestern United States is that the coal-producing facilities are far away from the coal consuming centers, thereby necessitating higher transportation costs. Alternate methods of coal transportation by unit trains and larger trucks and towboats have resulted in lowering these transportation costs. These have been supplemented by extra-high-voltage (EHV) transmission of coal-generated electricity from power plants at or near the mine site. Another attractive but water-consuming method is coal transport by slurry pipeline.

In recent years environmental considerations in coal mining have received equal importance and considerable emphasis is placed on reclaiming damaged lands and reducing air and water pollution resulting from mining and utilization of coal. These include reclamation of surface-mined lands, filling of dangerous voids and abandoned mines, control of mine fires, sealing to inhibit acid formation, more effective methods for solid waste disposal, and more efficient removal of sulfur dioxide and dust from flue gases.

Increased research and development activities in the coal industry have also taken place in recent years. These include coal carbonization, coal combustion, and processes for the conversion of coal to pipeline gas and to gasoline and organic chemicals. Other potential advances in technology include magnetohydrodynamic generation of electricity with coal-fired generators, power from fuel cells, and the technique of solvent extraction of coal followed by hydrogeneration of extract to produce a refinery feedstock and ultimately high octane gasoline.

Coal Gasification

Production of gaseous and liquid fuels from coal is technically feasible but is only approaching commercial viability at present in the United States (Fair et al. 1976). This technology is capital intensive and fraught with environmental concerns. Probably the major drawback to conversion of coal to gaseous or liquid fuels is the relatively large quantity of costly hydrogen required for such conversion. Several new schemes are currently under consideration for reducing the cost of producing the required hydrogen.

Of all the new energy schemes under investigation, the production of pipeline quality gas from coal appears most promising. It is evident that the large-scale commercial conversion of coal to gaseous and liquid fuels will depend on many other variables besides technology. These include the rate of new oil and gas discoveries, oil import policies, price of indigenous oil and gas supplies, and the progress made in processing oil shale.

Because of the inherent problems associated with surface gasification processes, an alternate approach is underground coal gasification. This latter technique requires no mining and no surface reactor and hence the environmental considerations and economics appear very attractive in comparison to conventional techniques requiring underground mining and extensive surface facilities. However, there are some disadvantages associated with underground gasification such as decreased operational control, higher heat losses and some environmental problems associated with *in situ* extraction.

The *in situ* coal gasification concept is not new; the concept was first proposed by the British in 1868 and subsequently by the Russians in 1888. In Russia this technique has been under continuous investigation since 1930. In the United States, the concept was first investigated by the USBM in the 1950's, followed by Gulf Research in 1968. Although the initial investigation indicated technical feasibility of the process, the economic viability did not appear to be attractive at the prices for fuels prevailing at that time.

Since the oil embargo of 1973 and especially the more recent indication of impending energy shortage and increased energy costs, the concept of *in situ* coal gasification has taken a prominent position as one of the most promising and potentially viable techniques for recovering the fuel value and by-product chemicals from deep-seated coal deposits.

Especially, *in situ* coal gasification appears to be ideally suitable for coal deposits located in the water-deficient areas of the Western United States. These deposits have thicker seams, are non-swelling, contain sulfur and are of lower rank. Most important of all, these Western deposits constitute more than two-thirds of the total coal reserves potentially amenable to *in situ* gasification in the United States.

Currently, four different coal gasification processes are available:

1. The linked vertical wells concept.

2. The packed bed reactor concept proposed by Lawrence Livermore Laboratory.

3. The longwall generator concept.

4. The ERDA concept for steep-dipping seams.

Of the above concepts, the first involving the linked vertical well appears to be the best process, resulting in 80 to 100 percent utilization of the coal as compared to 40 to 50 percent obtained in underground mining due to the necessity of leaving coal in columns or pillars to prevent subsidence and maintain mine integrity. Figure 7.2 shows the schematic of the Linked Vertical Wells concept of underground coal gasification (Garon 1976).

OIL SHALE

One of the most promising fuel mineral resources located in the water-deficient regions of the Western United States is oil shale contained in the Green River Formation. These oil shale deposits cover more than 16,000 contiguous square miles in the adjoining states of Utah, Colorado and Wyoming and contain over 2 trillion barrels of oil. Some of these formations of economic importance are 30 feet or more thick, are usually less than 1,500 feet below the surface and contain an average of 30 gallons of oil per ton. It is estimated that, depending upon the depth of these formations, about 40 to 60 percent of these deposits could probably be recovered. For this reason, considerable incentive has been shown in the

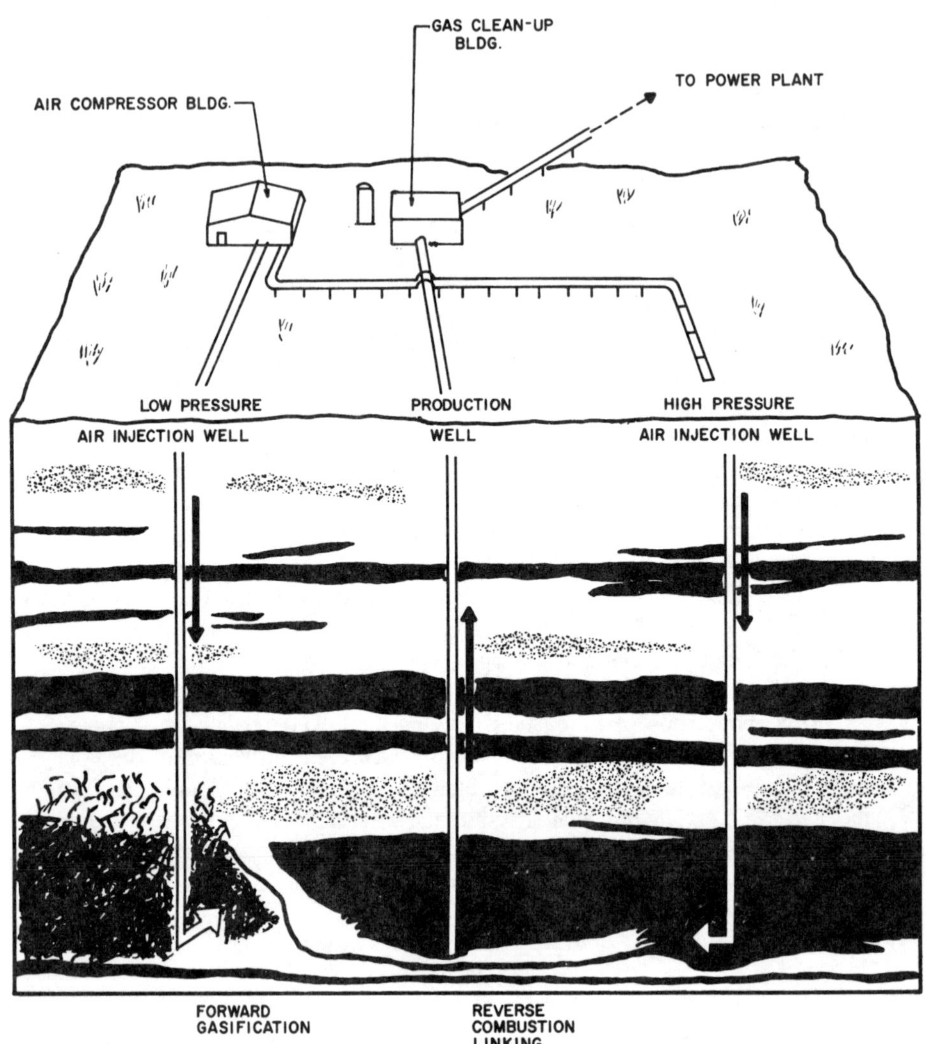

Figure 7.2 Linked vertical wells concept

development of the United States oil shale industry during the last few years (Prien 1977).

On the basis of the research and development work carried out by government agencies and the industry, an extraction sequence of underground mining (using the room and pillar method) and surface retorting has developed to produce oil from these shales. Retorting techniques with potential commercial scale-up include:

1. The TOSCO II and Lürgi-Ruhrgas processes which utilize recycled hot solids for heating.

2. The Paraho Direct Mode process which uses an internal combustion zone within the retort.

3. The Union Retort B., Superior Oil, and Paraho Indirect Mode processes which utilize external, fuel-fired furnaces as heat sources.

4. The Institute of Gas Technology (IGT) hydrogasification process capable of producing pipeline quality gas.

In Situ Retorting

In place retorting of oil shale has always been an attractive alternative to conventional mining and surface retorting of shale. Some of the major advantages of this *in situ* processing of oil shale include less materials handling, elimination of the necessity for disposing the spent shale (tailings), reduced capital and operating costs, less use of water (a very attractive benefit in water-deficient areas such as Western United States), and an environmentally more attractive situation than conventional techniques. *In situ* retorting of oil shale has been under serious investigation since 1944 both by private industry and the federal government.

Anvil Points, Colorado (1945-56) was the first of several programs for large scale experiments in oil shale mining authorized by Congress in 1944. Sinclair Oil and Gas Company (Colony Development), Equity Oil, Mobil, Humble, Union, Shell and Paraho have all been associated with the development of this promising technique. Between 1964 and 1968 a nuclear *in situ* experiment (Project Bronco) was proposed, but never carried out because of public protest.

The conventional approach to *in situ* processing of oil shale (currently designated as "true *in situ*") consists of drilling a pattern of wells from the surface into the deposit, followed by fracturing between wells by hydraulic pressure, chemical explosives or by using nuclear devices. Next the fractured oil shale is ignited and the hot combustion gases forced through the shale to convert the organic matter to oil which is pumped from the peripheral recovery wells.

In 1972, Occidental Oil Shale began field investigation of its vertical, modified *in situ* process, involving underground mining followed by in place retorting of the oil shale in Colorado. In this technique about 10 to 15 percent of the shale in the chimney is removed to provide void space. Next the oil shale is drilled and blasted to completely fill the chimney with broken (rubblized) shale for *in situ* retorting. The combustion gases resulting from the retorting operation are removed by appropriate piping and the liquefied oil is pumped from sumps provided at the bottom of the chimney. Figure 7.3 is a schematic of the Occidental's *in situ* mining and recovery scheme. Using this technique, an overall yield (recovery) of about 70 percent was predicted under *in situ* retorting conditions (Crookston 1976).

It should be noted that *in situ* extraction of oil from oil shales is still in its infancy and considerably more research and development and pilot plant testing (demonstration plant) are needed before this technique is ready for commercialization. However, the *in situ* technique has demonstrated its usefulness and attractiveness, especially in water-deficient areas and thus has a very promising future.

URANIUM

The uranium industry in the United States has undergone a remarkable series of transformations during the last 35 years; from a commodity of minor commercial interest to one vital for nuclear weapons as well as a source of fuel for electrical energy. The United States is endowed with considerable reserves of uranium ore, most of which are located in th water-deficient areas of the Western and Southwestern United States, especially the Colorado Plateau and the Wyoming Basins. The ore is located for the most part in sedimentary formations such as sandstones, mudstones and limestones.

Most of these uranium deposits are located within 800 feet of the surface and several of the shallower deposits (up to 300 feet or so) are mined by open pit methods while the deeper ones are worked by various underground mining techniques depending on the size, shape, depth, grade, water table and nature of the ground. These mining methods include room-and-pillar, longwall retreat, and panel methods. One of the major problems in underground mining is the ventilation due to a presence of radon and its radiation hazard.

The basic steps in uranium ore processing include: (1) ore preparation, crushing and grinding; (2) uranium dissolution by acid or carbonate-bicarbonate leaching; (3) solid-liquid separation either by countercurrent

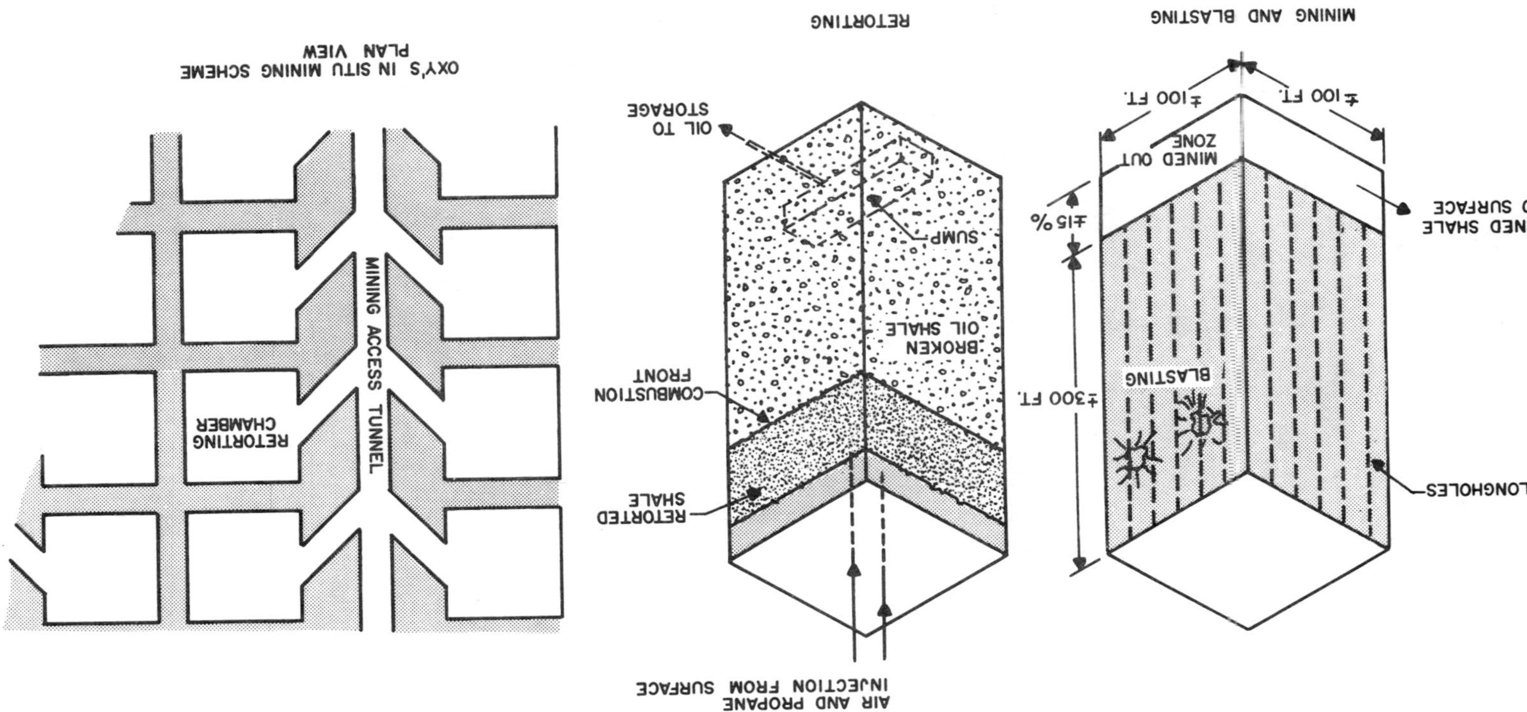

Figure 7.3 Occidental's *in situ* mining and recovery scheme

decantation or by filtration; (4) uranium concentration either by ion exchange or solvent extraction; and (5) yellow cake precipitation. The above conventional flowsheet is a hydrometallurgical one using water (with a water to solids ratio of 3:1) which, of course, is a major consideration in the arid regions of the Southwestern United States. However, due to utilization of optimum water recovery systems, more than 75 percent of the water is recycled for processing.

One of the major problems facing the uranium industry is the disposal of tailings resulting from the above hydrometallurgical processing. These tailings contain radon and other radionuclides and if the tailings are not properly impounded and sealed, there is the danger of radioactivity seeping into the surface or ground water and escaping into the air. Due to the long half life of the radionuclides, contamination of the tailings lasts practically forever.

The above hydrometallurgical process for uranium extraction is a very efficient technique resulting in more than 90 percent extraction of the uranium value contained in the ore. However, it is an expensive process and even at the current price of \$43.25 per pound U_3O_8 certain lower grade underground uranium reserves are considered submarginal. At the present time this cut-off grade is around 0.05 percent U_3O_8.

In Situ Extraction of Uranium

In situ, or solution mining is the in-place extraction of metals from ores located within the confines of a mine (unfractured or fractured ore, stope fill, caved material, and ores in permeable zones) or in dumps, ore heaps, slag piles, and tailing ponds. These materials represent an enormous, untapped, potential source of all types of metals. The field of *in situ* mining encompasses the preparation of ore for subsequent in-place leaching, the flow of solutions and ionic species through rock masses and within rock pores, the leaching of minerals with inexpensive and regenerable leaching reagents under conditions prevailing in-place, the generation and regeneration of such solutions, and the recovery of metals or metal compounds from the metal-bearing solutions. Accordingly, the overall scope of this potential mining method embraces interdisciplinary science and technology requiring application of principles of basic sciences, mineral technology, hydrology, and economics.

Advantages of *in situ* mining include environmental attractiveness due to less land disturbance, improved mineral utilization due to processing of lower and submarginal ores, and favorable economics due to earlier return on investment. Some of the disadvantages of the process are possible

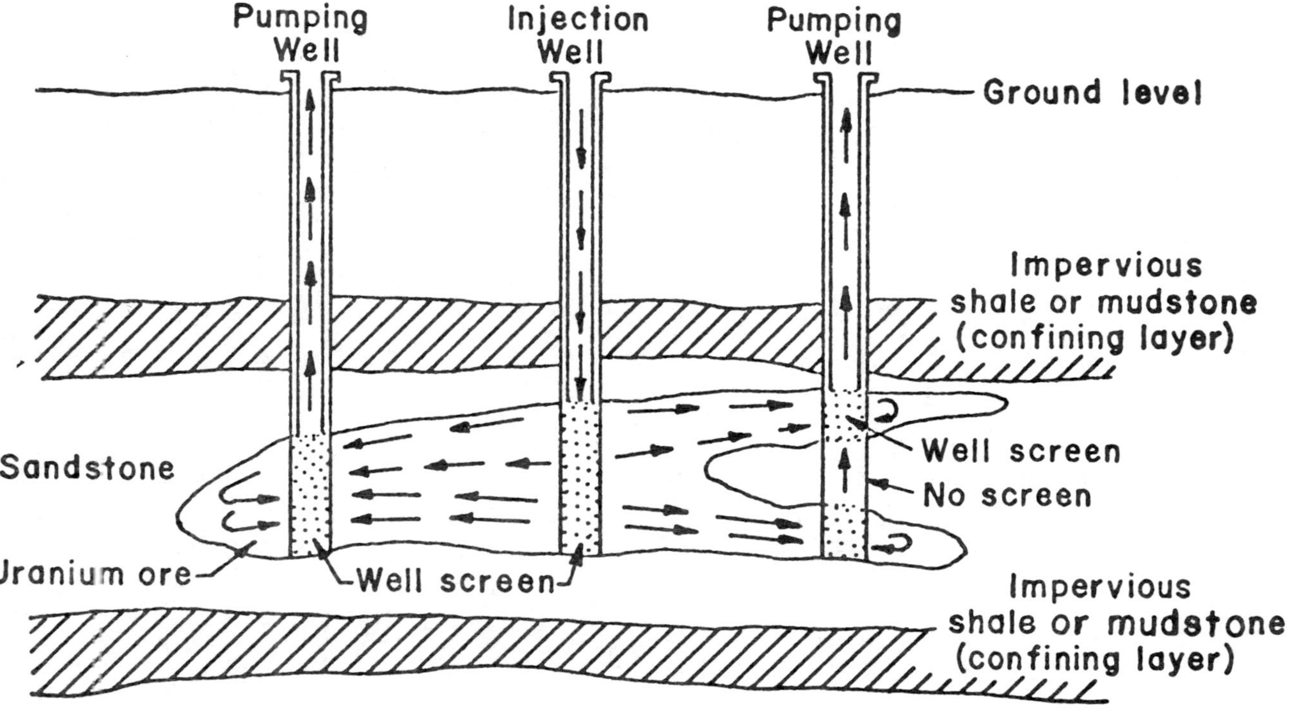

Figure 7.4 *In situ* leaching uranium-leach field

contamination of ground water, potential land subsidence, and lack of field and operational experience.

A detailed bibliography on *in situ* leaching technology has been compiled by Twin Cities Mining Research Center of the United States Bureau of Mines (Larson and D'Andrea 1976).

In situ extraction of uranium using the "bore hole mining" technique has proven to be one of the most economically successful techniques for recovering this metal from permeable sandstones, especially from lower grade deposits that cannot be mined and processed by conventional methods. This newly developed technique consists of a properly designed pattern of wells suitable for injection of the leach solution plus recovery of metal-bearing liquors through recovery wells. The wells are similar to water wells except that the casings are perforated only in the ore zone for optimum leaching conditions. The system includes monitor wells to detect any objectional movement of leach solutions in the ground water.

This *in situ* extraction technique is also very attractive in the extraction of uranium resources in the arid regions since very little water is needed for processing. Most of the process water is taken from the same or neighboring aquifer which contains the deposit and in actual practice about 3 to 5 percent excess water is pumped from the producing wells and this excess water needs to be disposed of. Thus there is a net gain rather than a loss of water in this technique.

In the evaluation of a uranium-mineralized block as a potential producer by in-place leaching using the bore hole mining technique, it is first necessary to establish a grade-thickness product which will cover development costs and operating costs with something left over for recovery of investment in plant and equipment for profit. This grade-thickness product is then a cutoff parameter for determining whether another hole is included in reserves or excluded.

Currently, there are more than 25 *in situ* mining operations recovering uranium from lower grade resources in this country. Most of these operations use a carbonate-bicarbonate leach solution in conjunction with a suitable ion exchange system for recovering the uranium values from leach solutions followed by elution of the leach resins and precipitation of yellow cake. By utilizing this technique, overall uranium extractions ranging from about 50 to nearly 100 percent have been reported. Moreover, economically, the *in situ* extraction is very attractive since the capital and operating costs required are much lower than for a conventional underground mining and surface processing plant. Also, it is possible to get into an operating position in a much shorter period than the conventional mining operation, thus resulting in a very favorable cash flow situation (Bhappu

1978). Figure 7.4 shows a schematic of the typical *in situ* uranium mining process.

Although *in situ* leaching in recent years has matured into a viable metal extraction process, considerable additional research is needed to bring it in competition with other high-confidence and universally acceptable processes.

The principal areas of research required to accomplish the above goal are:

1. Improved techniques for measuring physical properties of ores in place.

2. Improved techniques for drilling and blasting to produce desired fragmentation.

3. Improvement in injection and recovery systems.

4. Improved understanding of hydrology in broken ore masses.

5. Design of leaching systems for maximum oxidation.

6. Basic studies on leaching of dispersed mineral particles.

7. Improved methods of scaling up from laboratory tests.

8. Optimum parameters for leaching of massive deep-seated deposits.

9. Improved environmental controls.

REFERENCES

Agricola, G. (1950) "De Re Metallica," Hoover Edition, Dover Publications, Inc., New York, New York.

Bhappu, R. B. 1978. "In-Situ Mining Technology," *Underground Space,* 2:211-221.

Bray, J. L. 1947. "Non-Ferrous Production Metallurgy," John Wiley and Sons, Inc., New York, New York.

Cloud, P. E. Jr. [ed.] 1969. "Resources and Man" by Committee on Resources and Man, National Academy of Sciences—National Research Council, W. H. Freeman and Co., San Francisco, California.

Crookston, R. B. 1976. "Mining Oil Shale," presented at International Joint Petroleum-Mechanical Engineering Conference, Mexico City, September 19-24, 1976.

Deju, R., Bhappu, R., Evans, G. and Baez, A. 1972. "The Environment and Its Resources," Gordon and Breach, New York, New York.

Fair, J. C. et al. 1976. "An Evaluation of In Situ Coal Gasification," *Energy* 1:77-94.

Garon, A. M. 1976. "An Evaluation of Underground Coal Gasification," Second Annual Underground Coal Gasification Symposium, Morgantown, Pennsylvania, August, 1976.

Larson, W. C. and D'Andrea, D. V. 1976. "In Situ Leaching Bibliography, Parts I and II," U.S.B.M. (Part I available in updated form).

Prien, C. H. 1977. "Current Development in Oil Shale Processing, Energy and Mineral Resource Recovery," ANS Topical Meeting, DOE CONF-770440, April 12, 1977.

United Nations. 1968. "Proceedings of the UN Interregional Seminar on Ore Concentration in Water-Short Areas," United Nations Publication E.68.11.E.4, New York, New York.

8

INFRASTRUCTURE AND FINANCIAL REQUIREMENTS FOR ENERGY RESOURCE DEVELOPMENT

D. G. Nichols

Research Triangle Institute
Research Triangle Park, North Carolina

and

G. R. Stairs

Center for Resource & Environmental Policy Research
Duke University
Durham, North Carolina

SUMMARY

The focus of the chapter is on energy resource recovery development in the arid zone regions of Arizona, New Mexico, Colorado, and Utah. Reference is also made to other western states such as Montana or Wyoming where energy development impacts may be equally great. Issues of human resources and related institutional constraints are discussed briefly. Coal and oil shale development are given primary attention, although there are other potential energy opportunities in this region. The treatment progresses from social and socio-political considerations, through energy development potentials, to capital investment requirements for specific types of energy development.

INTRODUCTION

Arid lands in the United States may be variously defined geographically; however, the usual focus is upon the Southwest region. Here, in four states—Arizona, Colorado, New Mexico, and Utah—one finds the highly mineralized region characteristic of the mountainous west. The region is further character-ized by: (1) land ownership pattern strongly skewed to public lands; (2) a growing in-migration rate symptomatic of the demographic shift to the southern rim of the U.S.; (3) a region of diverse income levels, with a generally low per capita income average by national standards; (4) a fragile ecosystem in terms of vulnerability to man's transgressions; (5) a region where water is severely limited; and (6) widely dispersed major population centers with many small remote towns interspersed. While not the only arid or semiarid region in the U.S., this region serves as a major example for our purposes, and much of what is found here can be projected to similar, smaller regions of the arid zones.

The Southwest is also a region that has beckoned the adventuresome of western man since the early Spanish explorers. Gold and silver exploration and open-pit copper mining activities have been extensive. The cowboy frontier, well documented in fact and fiction, has been replaced by the mini-ranchette. The development of copper mining, and in some instances a closing-down of mining centers has tended to repeat the cycle. The rapid growth of an area based upon instant development is not always followed by equally rapid restoration when the development is over. Often a long protracted period of social marginalization takes place. The more aggressive depart, leaving a residual of under- or unemployment and urban decay of towns never finished.

On a more contemporary level the region may be characterized as one in which a new development thrust from population in-migration is about to be further accelerated by the search for and development of energy resources. Regional development has been under study by the "Southwest Region Under Stress Project," a cooperative research project among Southwestern researchers and other scientists in the country (Kneese 1978; Brown and Kneese 1978). A measure of "boom" seems evident in relation to energy resource development and must be dealt with by the region in foreground planning. Whether there will be corresponding "busts" remains to be seen—and guarded against.

HUMAN RESOURCES AND INSTITUTIONS

The population of the arid Southwest has its own unique qualities. A large Spanish speaking minority, the largest remaining blocks of native American population, and a heterogeneous income level that ranges below the national average. It is a population in which much of the poverty is hidden from planners and policy makers (Center for Public Affairs, 1977) and one that still lacks heavy investment in social programs. While it is beyond the terms of reference for this chapter to discuss the array of human needs in the region, two factors stand out as relevant to the energy development question. The first is the issue of articulation between low income people and the new employment market that may be created by an expanding energy development. The second is the status of energy development on Indian lands and its relation to the future of the native American people.

A fundamental infrastructure need is for region wide attention to the low income and minority people that could be greatly benefitted by a new employment structure. As noted, the poor in this region are often rural and small town based, lack advocacy representation and thus are difficult to view in the political sense. They are also often miscategorized as being related only to the agricultural and traditional rural sectors since these areas provide a minimum level of employment and generally require low skills.

An articulation of recognition, training and education programs, and employment opportunities is required to provide full development of the region's human resources. The economics and organizational structure necessary will require close cooperation between the private and public sectors. The socio-economic health of the region could be well served by placing policy emphasis upon utilizing energy resource recovery as a means for balanced socio-economic development rather than allowing a "boom-bust" process to take place.

The native American population in the region is of special interest relative to energy resource development for at least two major reasons. First, they represent a low income group almost without parallel in the U.S. Second, their lands hold by current estimate a significant fraction of the nation's coal and uranium reserves. Taken together those two points immediately suggest a solution—that energy resource development will provide a means to an end, that of self-sufficiency and a better life for the Indian people. Obtaining that solution may be more deceptive than it first appears for the Indian situation is fraught with many political, legal, social and economic constraints.

These issues have been well addressed in the recent American Indian Policy Review Commission Report (1977). Also, the Indian people are making their own firm plans for energy development. The Council of Energy Resource Tribes (CERT) was started in 1978 to provide technological and professional backup for energy development on Indian lands. In brief, the Indian position seems to be one of seeking to drive a hard but fair bargain on the development of their resources. They have added to their management abilities through CERT and also by direct creation of intra-tribal administrative structure for energy resource management. Wise attention is also being paid to environmental, social and cultural concerns, although much remains for confirmation.

The question of jurisdictional control is a paramount issue in Indian energy resource development. Given the strong desire for self-determination it would seem apparent that tribal government may prevail. Among issues left unanswered are the tribes' response to federal and state regulations as contrasted to or in addition thereof to their own policies. Since collective national reserves on Indian lands may represent nearly half of our coal and uranium supplies the matter is not a small one. At the present time the Indian lands provide less than 5 percent of the nation's total gas and oil output, but the move toward coal for energy could dramatically change the situation. The tribes' ability to tax and regulate extractive processes will affect their competitive posture vis a vis other resource development projects. The ability to vary taxation and royalties could provide the Indian with a competitive advantage in certain instances. For example, taking their financial returns in the form of taxes rather than royalties might allow the contractor to deduct the tax from his own Federal tax burden and thus make operation on Indian lands more attractive.

LAND AND NATURAL RESOURCES

The arid land region in the U.S. is coincident with a region of massive public land ownership. States in this area often range from 50 to 80 percent public lands with a dominance of federal lands. Land use decisions are thus, in theory, dependent upon an interaction between federal, state, and private sectors. In fact, however, most decisions in the past have usually been ownership specific with rather little coordination between ownership classes. In some instances there has been remarkably little interaction even between the branches of federal government charged with land management.

Issues of resource use on federal lands and in particular mining and nonrenewable resources extraction are the subject of much current study. Current mining laws are still referenced to a bygone era and possible adjustments in mining laws may be appropriate. An exception to the notion that land-use agreements have seldom included the various ownerships may be found in the grazing land leases of this region. Historically, local ranchers have depended upon a small portion of deeded land with access to large units of public domain on a lease base to sustain their operations. One can visualize a potentially similar situation developing in the coal or oil shale leasing program.

Land Use Problems

The institutional problems inherent in land usage include variations in lease policy and regulation, environmental constraints, access, taxation, and so forth. On federal lands one is forced to do business with a number of agencies, bureaus, or services housed in one of two departments: the Department of the Interior or the Department of Agriculture. In addition, agencies such as the Department of Energy or the Environmental Protection Agency are specifically involved in federal lands operations. States have begun to add their own set of environmental regulations and policies, and in some instances even local governments have added regulations. Thus a given operation may impinge upon several ownerships, be impacted by a variety of policies, and require access across lands all derived from differing ownerships and/or management philosophies.

Taxation of energy resource development operations raises a special concern. Major tax issues at the state level include service taxes, property taxes, license taxes, and increased personal income or sales taxes resulting from larger payrolls and expanded local economies. In addition, if conversion plants are built to export energy by wire or by gas pipeline, an additional tax of significance might be assessed against the production plant. The Navajo people have already proposed a pollution tax for such purpose and some states have, or have proposed, a tax on electrical generation (for example, New Mexico and Montana). Issues of taxation may be made more complex by the nature of ownerships and operations. The mill site may be in a local government district different from the extraction site. Workers may live in a community politically isolated from the actual production sites. Since the funds available from energy resource development taxation (including associated tax bases) may eclipse present public revenue sources, it is a significant factor for state and local planners. There is an evident need to program for the future and to provide an equalizing infrastructure to deal with inequities that may be created. And, there may be increasing concern about "windfall profits" from the rest of the nation if a tax burden placed upon national consumers becomes unduly burdensome.

Indian lands, while generally understood to be government-held trust lands, are regarded by the Indian people as the foundation of a sovereign nation. Thus, they can argue, with certain legal standing, that theirs is a land whose owners can choose to set their own legal structure. Obviously this attitude, whatever its legality, could create considerable consternation. Further, the magnitude of energy deposits on Indian lands emphasizes the urgency of resolving these issues. A general attitude that federal laws apply, but that state laws generally do not, has often been suggested. However, some states have been given greater authority through congressional transfer, the final validity of all these legal questions remains for determination. Nevertheless, the energy resource development on Indian lands must be treated as a special instance subject in the final analysis to tribal authority.

Water Resource Questions

If the lands situation seems complex, one must view the water resources of the arid region as an even greater problem. In a region where water is limited to much of the proposed development, it is the pivotal issue to be addressed. At present, water use in the arid West follows the legal doctrine of "first in use, first in right." Since agriculture was an initial thrust of development in the region, it is not surprising to find that 80 percent of the water is used by this industry. As energy resource development begins, the arguments about water use can only be increased. At the present time, there are at least four major interest groups concerned with water use: the mines, prinicpally copper mining; the agriculturists; the Indian people; and the general public. The federal government is also involved, primarily through major water projects and the "implied reservation doctrine." The latter is a legal question of great importance to both federal lands and Indian lands. In summary, the doctrine suggests that at the time of initial designation of federal and Indian lands, there was reserved an adequate water supply for all the arable land and its development. Taken to its most logical conclusion, this doctrine effectively precludes other water development, and in fact, has already been violated by current development. Clearly these issues must be ranked prior to new allocation of water or its redistribution.

Energy resource development will place increasing pressure upon an already limited supply of water. New infrastructure will be required to provide even compromise solutions. Although conservation and prioritization of uses may help, one may expect that a natural consequence of absolute restricted supplies will be felt primarily in economic terms. The ability of current rights-holders to sell water rights could become a new major industry in the arid regions. Certainly water availability and cost need to be programmed carefully into future development planning.

Coal and Oil Shale Demand/Supply Considerations

The reserve base of coal in the western states indicates 167 billion tons with a sulfur content less than 1 percent, 37 billion in the 1.1 to 3.0 percent range and a total of 234 billion tons. (The estimated U.S. total is 436 billion tons). Projections have placed the demand for this coal at some 1350 million tons per year in the 1985 to 1990 time period; of this total, 880 million tons would be consumed by electric utility companies, 43 million tons[1] used for the production of synthetic fuels, with 427 million tons going to other uses. Overall, this would represent a substantial increase over the current level of production. And, most of the increase has been projected for the western states. Average growth rates of 10 percent per year for the West and 3.5 percent per year for the East are typical of many projections for the time period of 1980 to 2000 (c.f., Congressional Research Service, 1978; Gustaffero, et al., 1977; Griffith and Clark, 1979).

The National Energy Plan (NEP) has projected that coal production by 1985 can grow to 36 million tons per year in the Southwest under the influence of the market place, or to 65 million with the added stimulation of a federal program to promote coal utilization, that is, NEP. The corresponding values for the east-north Great Plains are 220 versus 226 million tons per year. For the western U.S. these values are 360 versus 434 million tons per year, and constitute some 34 percent of the nation's total projected coal production.

A substantial increase in the available transport system for coal will be needed since, unlike oil shale, it is often shipped great distances before being used (converted). Three primary modes for western coal transport are seen. Railroads which mostly support unit train coal movement provide the most flexible means. However, for the long distance transport of large quantities of coal between the same terminal points, pipelines which deliver coal in the form of a water or methanol (methyl alcohol) slurry may be desirable. The Black Mesa pipeline transports Arizona coal over 273 miles of rough terrain to an electric power plant. Nichols and Chuang, 1976, have examined various transport systems associated with coal utilization. And, the study of Bilsky and Leesley, 1979, specifically compares the transport via railroad with pipeline systems involving water or methanol slurry. These latter alternatives are shown to possess cost advantages under particular circumstances including the availability of water or methanol. The methanol would be produced on a large scale via coal gasification to produce synthesis gas, that is, a mixture of carbon monoxide and hydrogen, which can be reacted over appropriate catalysts to form methanol. Such gasification plants also require substantial water resources.

Oil shale reserves have been rated in terms of the technology currently being commercialized for its conversion to a synthetic crude oil. Recoverable under present conditions are some 80 billion barrels of such oil, from shales yielding 25 to 100 gallons per ton of raw material. This resource is located in Colorado, Utah, and Wyoming. (An additional 500 to 2000 billion barrels of marginally to submarginally recoverable oil may be available, depending upon demands and prices).

The major oil shale development activities of the U.S. are described in a variety of sources (c.f., Berry, 1979; Congressional Research Service, 1978; Staff Report (DOE), 1979). A number of advantages have been described for shale conversion facilities which incorporate both *in situ* (underground) and surface retorts. These advantages include higher oil yields, less water requirements, and significantly reduced environmental disturbance from spent shale disposal. These factors result in lower costs.

REGIONAL DEVELOPMENT

Increased coal production in the West, and particularly in the states of the Southwest will involve large surface mining operations. Little or no coal cleaning will be necessary; thus, the land requirements for mining will be relatively large while the water needs will be small. Railroad transport to conversion plants is likely to be the predominant mode of haulage with trucks being involved in short distance movements. Slurry pipeline systems are severely constrained; essentially no infrastructure exists for such systems and they require the dedication of very large reserves to specific suppliers and users. Long distances must be involved to justify such ventures and so large capital outlays are necessitated.

It can be expected that both the coal and oil shale of the West will go to serve population areas far distant from the fossil energy source. Coal will in more cases be transported near to consumption centers before conversion to other energy forms. The shale will be converted near its source due to the large weight ratio involved, i.e., shale mass/shale oil mass. Conversion in the case of coal will be to electricity via coal combustion and steam turbine units and to fuel gases via steam/oxygen gasification reactions to form a medium heating value product gas. This product gas could be used locally as an industrial fuel gas; alternatively, it can also be processed and upgraded for use as a chemical feedstock, for example, methanol production from synthesis gas, or as a source of SNG (synthetic natural gas). Other uses for the coal do not appear to be promising due either to the lack of commercially available processes or the large distances to consumption centers.

The shale will be retored in both underground (*in situ*) and surface heating chambers or retorts, the product(s) will include a synthetic crude oil, some low quality fuel gases, and possibly a few bulk chemicals. The synthetic crude oil will necessitate special processing or blending operations when upgraded to a commercial heating oil or engine fuel. The cost of these additional processing steps, while smaller than that for the shale retorting plant operations has not been fully determined. An overview of the U.S. goverment objectives in promoting the commercialization of new energy conversion technologies has been prepared by the Department of Energy, 1978.

Coal Combustion or Gasification

The primary use for coal in the U.S. has been for electricity production via combustion to generate steam for turbine/generator sets. (Combined cycle processes are under development in which coal is gasified to a fuel gas for feed to combustion turbines which function in association with steam turbines. While such systems enjoy an overall higher thermal efficiency, high costs and a lack of commercially available combustion gas turbines prevent the use of this option between now and 1990). Conventional pulverized fuel boiler systems are reliable and in wide use currently. These will clearly receive future use; however, boiler modifications are being made to provide reduced emissions of nitrogen oxides, electrostatic precipitators or baghouses appropriate for the control of particulate emissions, and flue gas desulfurization is to be mandated so that sulfur oxides emissions will be quite low. Fluidized bed combustion boiler units are now at an advanced stage of development. Electric power plants based on this approach may also be constructed, although the tradeoffs of technology costs versus environmental control benefits make this option more viable in the eastern U.S. where the coals are generally higher in sulfur content.

Data on the costs of electricity generation via coal combustion as well as on costs for SNG production from coal have been compiled and compared at the Institute of Gas Technology and the Gas Research Institute (Linden, 1979). It has been shown that for plants producing the same energy output that coal gasification involves about 1/2 the capital investment and about 1/3 the delivered residential energy price as that for electricity production. Table 8.1 shows such a comparison; here it is assumed that both installations would require all new facilities with appropriate environmental control equipment, i.e., scrubbers, etc., and that the energy product is to be consumed a distance of 1000 miles from the plant site. The fact that coal gasification is a viable alternative form of energy conversion thus is clearly revealed.

Coal gasification plants are already being operated on a commercial basis in various other countries. These are based primarily upon the fixed bed or atmospheric entrained-phase mode of reaction. The Lürgi fixed bed approach is well known and reliable for nonagglomerating coals, for example, western coals.

The costs associated with a coal gasification plant based on the use of Lürgi gasifiers have been estimated by Detman, 1976 and 1978, and compared with those for plants which would use alternative "second-generation" technology, namely, Bigas, Hygas, or Synthane gasifiers. These cost estimates are shown in Table 8.2. The capital investment required for a Synthane plant was higher than the others examined because byproduct char is produced in this alternative which must be accommodated; the char was assumed to be burned on-site in an electric power production facility, the electricity being a by-product to be sold at $0.004/kwh. The product SNG was estimated to have an overall production cost of $3.30 versus $2.71 per million Btu for the Lürgi and Hygas alternatives, respectively. The latter was the lowest of the estimated gas costs, expressed on a 1976 cost year basis.

The costs associated with the use of Lürgi gasifiers and the additional processing equipment needed for processing and methanol and/or gasoline production have been estimated by Schreiner, 1977. A 50,000 barrel per day methanol plant would require 1.6 billion dollars (1976). Gasoline, via Fischer-Tropsch synthesis using Lürgi gasifiers, was estimated to cost about four times that for methanol, on a volumetric basis, i.e., $1.67 versus $0.48 per gallon, as shown in Table 8.3. A number of alternative processes for methanol synthesis are commercially available. Gasoline synthesis processes are currently under development by Mobil and others. (Methanol could be produced near coal mines and used as a slurry medium to transport raw coal to market areas, as discussed earlier in this chapter).

Table 8.1. SNG versus electricity (delivered residential energy prices)

	Coal Gasification	Coal Electricity with Scrubbers
Capacity	250 million CF/day	3000MW(e)
Capital cost, 1977 $billion	1.4	2.9
	1977 $/million Btu	
Capital charge	2.51	8.09
Fuel cost	0.71	1.28
Operation & maintenance	1.20	0.86
Credit for byproducts	(0.56)	NA
Transmission & distribution	1.50	5.01
Total	5.36	15.24

Source: H. R. Linden. "Perspectives on U.S. and World Energy Problems." Gas Research Institute. February 1979.

Table 8.2. Costs for alternative coal gasification processes for synthetic natural gas production (cost basis:1976)

Process	Hygas	Bigas	Synthane	Lürgi
Plant equipment (million $)	870	1030	1330	1060
Capital requirement (million $)	1060	1260	1615	1310
Raw materials (million $/yr)	50.4	55.9	68.7	73.3
Gross operation cost (million $/yr)	123.6	141.7	166.2	167.8
Byproduct credits (million $/yr)	28.9	5.0	22.9	54.3
Net operating cost (million $/yr)	94.7	136.7	143.4	113.6
Average Gas Cost[a] ($/million Btu)	2.71	3.50	4.11	3.30
Constant Gas Cost[b] ($/million Btu)	3.91	4.93	5.93	4.78

[a]Utility financing.
[b]Private financing.

Source: R. Detman. "Factored Estimates for Western Coal Commercial Concepts." Interim Report. FE-2240-5. C. F. Braun & Co., Alhambra, California, October 1976.

Shale Oil Production

A number of studies have been conducted on the commercialization potential and costs of producing a synthetic crude oil from oil shale (see, Berry, 1979, Congressional Research Service, 1978; Pelofsky, 1977; Pforzheimer, 1978). A (small) plant using one surface retort could produce 4,500 barrels per day of crude product. A capital investment of some 65 million dollars and an operating cost of 9 million dollars per year would represent an overall oil cost of $19.00 per barrel (1976 dollars).

Some 24 of these 40 ft-diameter retorts would be needed in a plant having a capacity of 100,000 barrels per day. The 1.2 billion dollar capital investment, with a 150 million dollar operating cost per year, would represent an oil cost of about $12.00 per barrel (1976 dollars).

At some geographical locations underground retorting may be feasible. A number of advantages would accrue to a production complex involving simultaneous *in situ* and surface retorting operations. However, firm cost

Table 8.3. Gasoline production alternatives via coal gasification

Case	Coal MT/D	SNG MMSCF/D	Gasoline MBbl/D	MeOH MBbl/D	Efficiency HHV%	Operating Cost MM$/Y	Investment MM$	DCF Mode(C)		Utility(C)	
								Thermal(B) ¢/gal	Multiple ¢/gal	Thermal(B) ¢/gal	Multiple ¢/gal
I	27.3	148.5	22.0	—	62	132	1,732	85	117	62	85
IA	27.3	146.4	1.3	50.8	66	123	1,602	40	48	29	35
IB	27.4	—	37.4	—	47	149	1,975	121	123	87	89
IC	27.3	152.6	23.0	—	63	129	1,680	83	114	60	83
II	27.8	173.3	13.6(A)	—	58	152	1,886	93	167	69	122
IIA	27.3	177.9	17.5	—	60	135	1,748	86	142	63	103

[a]Also 3.4 MBbl/D of diesel fuel, fuel oil, and alcohols.
[b]At 5.1 MMBtu/bbl.
[c]DCF = Discounted-cash flow cost accounting procedure.
Utility = Utility cost accounting procedure.

Note: It was assumed that Lürgi coal gasification was used in each case. And, cases I, IA, IB, and IC involved gas processing followed by a conventional methanol synthesis loop.

Source: M. Schreiner, "Research Guidance Studies to Asssess Gasoline from Coal by Methanol-to-Gasoline and Sasol-Type Fischer-Tropsch Technologies," FE-2447-12. Mobil Research & Development Corporation, Princeton, New Jersey. December 1977.

Legend:
I. Methanol to SNG & gasoline (Mobil)
IA. Methanol itself as the product
IB. Methanol to gasoline (Mobil)
IC. Methanol to SNG & gasoline (fluid-bed)
II. Fischer-Tropsch direct to gasoline, etc.
IIA. Mobil direct process to gasoline, etc.)

estimates cannot be made at this time since this is a rapidly developing technological area. Berry (1979) has described many of the important considerations.

Projected Synfuel Capacities and Capital Requirements

For the purposes of this study, estimates were made for the magnitude of a synthetic fuel industry in the United States which could provide one million equivalent-energy barrels of oil per day of output from coal and oil shale. As shown in Table 8.4, this would involve coal gasification to produce industrial fuel gases (IFG), synthesis gas for further conversion to liquid fuels, and synthetic natural gas (SNG). No shale oil production is projected for the eastern or central regions of the U.S.; but, a "conservative" estimate of 150,000 barrels per day is projected for 1990 in the West. This estimate is conservative relative to the 400,000 barrel per day goal which was initially set in the President's major synfuel industry program plan.

Specific capital cost estimates recently developed as a part of the U.S. Department of Energy planning process were used to determine the level of overall capital requirement for synthetic fuels from coal in the west. Table 8.5 indicates that some 14.4 billion dollars would be needed to

Table 8.4. Projected synfuels capacity from coal and oil shale by 1990 for the United States

Fuel Type	No. Plants(A)	East	Central	West	U.S.
		Capacity[a]			
Liquids (coal)	20	250	150	100	500
Gas (SNG, coal)	8	62	62	126	250
Gas (IFG, coal)	30	60	60	30	150
Liquids (shale)	6	0	0	150	150
Total[b] (daily barrels, thousands)		372	272	406	1050

[a]Individual plant capacities may vary from 30 billion to 300 billion Btu per day output, for example.
[b]Thousand barrel per day crude oil equivalent at 6 million Btu per barrel.
SNG = Synthetic natural gas.
IFG = industrial fuel gas and/or synthesis gas.

Table 8.5. Capital requirements for projected synfuel industry using coal and oil shale by 1990 for western United States

Fuel Type	Capacity in West (thousand bbl per day)	Capital Cost[a] $/bbl[b]	Investment Estimate (million $)
Liquids (coal)	100	$48,000	4,800
Gas (SNG, coal)	126	$40,000	5,040
Gas (IFG, coal)	30	$27,000	810
Liquids (shale)	150	$25,000	3,750
	406		14,400

[a]Cost basis: 1979 dollars.
[b]Daily barrels of crude oil (equivalent) were used with each barrel representing 6 million Btu of heat equivalent energy.

finance the energy-equivalent of 400,000 oil-barrels per day of capacity. Of this, some 5 billions could be attributed to SNG production in the amount of 750 million standard cubic feet per day of high heating value gas, that is, three large scale gasification plants for SNG production.

One-hundred thousand oil barrels per day of energy-equivalent liquid fuels via indirect liquefaction, that is, gasification and synthesis, is projected at a cost of 4.8 billion dollars. The shale oil product can be produced more cheaply; though more detailed considerations should be applied for comparative purposes since the product quality and the environmental costs to society will be different per equivalent-energy product barrel for shale oil and for liquid fuels via coal gasification. Generally, the latter can be used as engine fuels while the former may be processed to a heating oil or further processed to engine fuel quality.

ENVIRONMENTAL AND ECONOMIC IMPACTS

A number of highly relevant studies have been performed on the environmental and economic impacts which may result from the construction and operation of a major synthetic fuels industry in the western coal and oil shale resource areas. The regional development impact studies of Radian (1975) and White (1977) have been comprehensive. The National Coal Utilization Assessment (1978) is a useful study and is presented in a systematic framework so as to delineate the various (potential) constraints on coal use in the various regions of the country. Siting problems, pollution effects, health effects and socioeconomic problems are described. Krutilla and Fisher (1978) have given special consideration to economic and fiscal impacts of coal development in the Northern Great Plains.

The National Academy of Sciences (1974) and Wright (1978) present data and discussion on the reclamation of western lands after disturbances due to mining and energy conversion uses. It is generally agreed that preplanning is required to properly assess and assign the costs associated with major land use projects in this region. Pressey (1979) provides a summary of environmental research on oil shale utilization.

The environmental impacts of two energy equivalent plants (see Table 8.1) are presented in Table 8.6. Note that the discharges (solid wastes and atmospheric pollutant emissions) from a coal gasification plant will be much smaller than those from a conventional coal-fired electric utility plant at the same site and using the same fuel. It is assumed that stack gas scrubbers are used on the combustion process waste gases to remove sulfur oxides. Also, note that the water requirement for the coal gasification plant is about 1/5 that of the electric utility plant, according to Linden, et al. (1979).

The unit product costs for energy via coal gasification, direct coal liquefaction, or shale oil production have been estimated (see Staff Report [DOE], 1978). These are shown in Table 8.7, along with that portion of the energy cost attributable to environmental control costs. The influence of stringent environmental standards can be seen to be a significant but not primary factor in determining total energy costs. Industrial fuel gas (low and medium Btu gas) and shale oil (surface retorts) are seen to be the most economical energy forms available from the coal and oil shale resources of the country. Careful planning of synthetic fuels and electricity production plants is essential to ensure technical operability, cost effective investments, and environmentally acceptable installations.

Table 8.6. Environmental impact of two equivalent energy projects

	Coal Gasification to SNG	Coal Combustion for Electricity
Unit Plant Capacity	250 million CF/day	3000 MW(e)
Discharge to atmosphere, lb/hr		
Particulates	180	870
Sulfur dioxide	450	2,300
Nitrogen oxides	1,780	20,830
Water requirements, acre-ft/yr	6,300	41,400
Solid wastes, tons/day	1,400	5,100

Source: H. R. Linden. "Perspectives on U.S. and World Energy Problems." Gas Research Institute. February 1979.

Table 8.7. Estimates of energy and environmental costs for synthetic fuels (cost basis: 1978 dollars)

	Current Standards		Future Standards (Most Stringent)	
	Total Energy Cost	Env. Cost	Total Energy Cost	Env. Cost
I. Coal Gasification				
High Btu	$4.6/MM Btu	$0.70/MM Btu	$5.1/MM Btu	$1.2/MM Btu
Low Btu	$3.6/MM Btu	$1.0/MM Btu	$4.1/MM Btu	$1.5/MM Btu
II. Coal Liquefaction				
(H-Coal)	$26/bbl	$2.9/bbl	$32/bbl	$9.3/bbl
	($4.5/MM Btu)	($0.5/MM Btu)	($5.6/MM Btu)	($1.6/MM Btu)
Surface retort	$18/bbl	$5/bbl	$24/bbl	$11/bbl
	($3.1/MM Btu)	($0.9/MM Btu)	($4.2/MM Btu)	($1.9/MM Btu)
In-situ	$25/bbl	$5/bbl	$33/bbl	$13/bbl
	($4.4/MM Btu)	($0.9/MM Btu)	($5.8/MM Btu)	($2.3/MM Btu)

Source: Staff Report (DOE), 1978. "Environmental Readiness Document(s)," Commercialization Phase III Planning, U.S. Department of Energy, Washington, D.C.

NOTES

1. It has been estimated that by the year 2000 some 100 to 700 million tons per year of coal will be consumed for synthetic fuels and related chemical feedstocks (c.f., Mann and Heller, 1979). Such quantities could provide 200,000 to 3,500,000 equivalent barrels per day of fuels. Specific projections are enumerated in the next section of this chapter.

REFERENCES

Bankers Trust Company. 1978. *U.S. Energy and Capital: A Forecast 1978-1982.*

Berry, K. L., 1979. "Combined Retorting Technique for Shale Oil," *Chem. Engr. Progr.* 75 (9):72-77.

Bilsky, I. L., and M. E. Leesley. 1979. "A Preliminary Evaluation of Methacoal as an Alternative Method of Coal Transport," *J. Air Pollution Control Assoc.* 29 (5).

Brown, L., and A. V. Kneese. 1978. "The Southwest: A Region Under Stress." *The American Economic Review,* 68, (2):105-109.

Center for Public Affairs. 1977. "The Rural Poor: Unseen by Policy Makers." Arizona State University, Tempe, Arizona.

Comptroller General, 1977. "First Federal Attempt to Demonstrate a Synthetic Fossil Energy Technology—A Failure," U.S. General Accounting Office, Washington, D.C.

Congressional Research Service. 1978. *U.S. Energy Demand and Supply: 1976-1985.* Final Report. Prepared for Subcommittee on Energy and Power.

Congressional Research Service. 1978. *Energy and the Economy,* U.S. Government Printing Office.

Department of Energy. 1978. "Overview of Technology Commercialization Assessment," DOE/US-003.

Detman, R. 1976. *Factored Estimates for Western Coal Commercial Concepts,* Interim Report, C. F. Braun & Company, Alhambra, California, 122p.

Detman, R. 1978. *Factored Estimates for Lignite Commercial Concepts,* Interim Report, FE-2240-98. C. F. Braun & Company, Alhambra, California. Prepared for the U.S. Department of Energy & Gas Research Institute.

Griffith, E. D., and A. W. Clarke. 1979. "World Coal Production." *Scientific American,* 240 (1).

Gustaffero, J. F., M. Maher, and R. Wing. 1977. *Forecast of Likely U.S. Energy Supply*/Demand Balances for 1985 and 2000 and Implications for U.S. Energy Policy, NTIS PB 266 240, Washington, D.C.

Hissam, M., and D. Nichols. 1976. "An Integrated Process Model of the Fischer-Tropsch Process for Liquid Fuels Production from Coal," *Proc. Seventh Pittsburgh Modeling and Simulation Conference.*

Kneese, A. V. 1978. "Status Report Southwest Region Under Stress Project." *New Mexico Business,* 31 (5):3-17.

Krutilla, J. V., and A. C. Fisher. 1978. *Economic and Fiscal Impacts of Coal Development.* John Hopkins University Press, Baltimore, Maryland.

Linden, H. R. 1979. "Perspectives on U.S. and World Energy Problems," Gas Research Institute.

Mann, C. E., and J. N. Heller. 1979. *Coal and Profitability: An Investor's Guide,* McGraw-Hill Publications Company, New York.

National Academy of Sciences. 1974. "Rehabilitation Potential of Western Coal Lands," Ballinger Publishing Company, Cambridge, Massachusetts.

National Aeronautics and Space Administration. 1977. *Energy Conversion Alternatives Study (ECAS)*, Summary Report. NASA TM-73871. Prepared for Energy Research & Development Administration.

National Coal Utilization Assessment. 1978. *An Assessment of National Consequences of Increased Coal Utilization*, Executive Summary. Prepared for U.S. Department of Energy.

Nichols, D. C., and D. C. Chuang. 1976. "Alternative Coal and Product Transportation Modes in Relation to Coal Conversion Processes," *Proc. NCA/BCR Coal Conference on Coal Utilization*, Louisville, Kentucky.

Pelofsky, A. H. 1977. *Synthetic Fuels Processing*, Marcel Dekker, Inc., New York.

Pforzheimer, H. 1978. "Oil Shale Comments." Rocky Mountain Oil and Gas Association Meeting, Denver, Colorado.

Pressey, R. E. 1979. *EPA Program Status Report: Oil Shale 1979 Update*, U.S. Environmental Protection Agency, EPA-600/7-79-089.

Radian Corporation. 1975. *A Western Regional Energy Development Study, Primary Environmental Impacts, Volume I*. NTIS PB 246 264. Prepared for Council on Environmental Quality and Federal Energy Administration.

Schmidt, R. A. 1979. *Coal in America: An Encyclopedia of Reserves, Production and Use*. McGraw-Hill Publications Company, New York.

Schreiner, M. 1977. "Research Guidance Studies to Assess Gasoline from Coal by Methanol-to-Gasoline and Sasol-Type Fischer-Tropsch Technologies," FE-2447-12. Mobil Research & Development Corporation, Princeton, New Jersey.

Staff Report (DOE), 1978. "Environmental Readiness Document(s)," Commercialization Phase III Planning, U.S. Department of Energy, Washington, D.C., DOE/ERD-0012, 0015, and 0016.

Staff Report (DOE), 1979. Oil Shale R, D&D Program Management Plan (Draft), U.S. Department of Energy, Washington, D.C.

U.S. Government. 1977. *American Indian Policy Review Commission Report*, Final Report. Vol.1-2.

White, I. L., et al., 1977. "Energy from the West: A Progress Report of a Technology Assessment of Western Energy Resource Development" EPA-600/7-77-072a-d (4 volumes), EPA-600/7-77-072a, EPA-600/7-77-072b, EPA-600/7-77-072c, and EPA-600/7-77-072d.

Wright, R. A. 1978. *The Reclamation of Disturbed Arid Lands*. University of New Mexico Press, Albuquerque, New Mexico.